Killing The Rumours

By John Charles Tyson

Copyright © 2021 by John Charles Tyson.

All rights reserved. No part of this publication may be reproduced, distributed or transmitted in any form or by any means, including photocopying, recording, or other electronic or mechanical methods, without the prior written permission of the publisher, except in the case of brief quotations embodied in critical reviews and certain other non-commercial uses permitted by copyright law. For permission requests, write to the publisher, addressed "Attention: Permissions Coordinator," at the address below.

John Charles Tyson C/- Intertype
Unit 45, 125 Highbury Road
BURWOOD VIC 3125
www.intertype.com.au

Ordering Information:
Quantity sales. Special discounts are available on quantity purchases by corporations, associations, and others. For details, contact the "Special Sales Department" at the address above.

Killing The Rumours/John Charles Tyson —1st ed.
ISBN 978-0-6452580-0-4

Contents

Acknowledgements 3

Prologue 5

Parameters 14

Articles to the first and only formal interview of James Tyson 15

The only formal interview of James Tyson on 8th Dec 1893 33

Articles from the interview to his death, 4th Dec 1898 54

Articles from his death to the trial to determine his domicile Nov 1901 84

Statements by witnesses in the trial to determine his domicile 108

Articles after the trial to determine domicile 162

RUMOUR: James Tyson never left Australia 183

RUMOUR: James knew his mother was a convict 198

RUMOUR: Tinnenburra woolshed was the largest in Aust 210

RUMOUR: James Tyson was a poet 215

RUMOUR: James Tyson never had any formal schooling 217

RUMOUR: William Tyson worked his fare to Aust through Mr Hartley 224

Queensland Business Leaders Hall Of Fame 229

James Tyson's Special Talents 231

Summation 237

Appendix 1 246

Appendix 2 247

Acknowledgements

Firstly, I must thank National Library of Australia, Digitised Newspaper section of Trove, because without that database this book would not have been possible. I spent many hours over several years searching and correcting the database about my Great-Uncle James Tyson.

My wife and rock, Libby, was always with me throughout the whole process. Many thanks Libby for putting up with me especially when things went wrong.

I thank Margaret Slatter and Zita Denholm for pointing me in the right direction in the beginning when I wasn't sure what I was doing.

The comments by my grandson, Matt Eddy, in the early versions were a great help in getting to the final product.

Many thanks to my cousins, Robert and Russell Jones, who were inspirational when assisting in getting the book to the "print ready" stage.

This book originated by default as I was researching for a novel about Uncle James and family. I was struck with the writings by people who knew James Tyson being vastly different to comments circulating in the community. I had to correct perceptions.

Now this book is finished I can return to the novel.

Prologue

I grew up during the Second World War. Japan bombed Darwin when I was two years old as the first and only aerial bombing of Australia. I remember when I was five seeing the search lights on the St Kilda foreshore; like silent sentries. But I believe never used. I was not frightened.

My first memory was living in one room of a rooming house in St Kilda, a suburb of Melbourne, with a shared bathroom. When I was about five, we managed to get a small kitchen down the passage which contained an ice-chest, gas stove, small cupboard, 2-person table and small bench with no sink. Two people could not work in the small floor-space left which gives an idea of the size. To get gas for an hour, sixpence had to be inserted in a meter in the passage.

A little later, an unlined sleep-out in the back yard affording little protection from the elements was rented. My brother and I slept there and my parents stayed inside the rooming house.

Dad worked in administration for the Air Force. He would leave very early in the morning and return when it was close to bed time. Fridays were good because that was pay-day and Dad would give my brother and me a packet of Steam Rollers, a peppermint flavoured sweet in a roll like life savers. However, I still remember over 70 years later the disappointment when one day he said no more because they are bad for teeth. And they were not replaced with anything else!

Mum had to learn dress-making so she could make clothes for other ladies and add to the family income. She had to see customers in that one room juggled around meals, homework and Dad's demands. My brother and I had to vacate when she had customers there. Now I don't know how she coped.

When I was eight, Dad went to Mildura, in northern Victoria, to establish a fifty-acre farming allotment in preparation for us all to join him. At this time Mum's dress-making produced the only income for this small family in St Kilda. No wonder Mum exploded and belted us for spending the sixpence we each had for Sunday School on buying rope!

It was soon after this time I heard about my rich uncle, James Tyson, how he had so much land it was calculated in square miles rather than acres. Even though I had the same surname, James Tyson seemed so remote from me and we were not part of the same family.

I put him at the back of my mind – until I finished full-time work. Through the magic of the internet, I found there was a plethora of information available about Uncle James. I was keen to find the real story of James Tyson; clearly the richest person in Australia in his time. He was born into an impoverished family yet shrugged off this burden and the lack of a formal education to amass his fortune.

I had heard rumours in my little circle that he was miserly, dishonest, uncommunicative and all he was interested in was making money. I was led to believe that he became so rich by claiming stray cattle where-ever he found them, particularly if they were unbranded.

A study of the available information about James Tyson suggested that it was unanimous amongst those who really knew him that the above rumours were untrue. However, people who heard about him or had a brief contact with him had been shown to think one or more of the above rumours were true.

It seems even his contemporaries, who believed they knew him, had little knowledge of who and what James Tyson was according to The Herald.

The Herald, Melbourne, Tuesday, 6 December 1898, P 4.

"JIMMY" TYSON. A REMARKABLE MAN JOINS THE MAJORITY. SOMETHING ABOUT HIM.

The sudden, though not altogether unexpected, death of Mr James Tyson, at the age of 76[1] years, has removed possibly the most picturesque figure in the history of Australia. Born near Sydney in 1823[2], the late Mr Tyson might well have been claimed by the Australian Natives' Association as one of the oldest and most representative members of the Greater Britain over-seas. That no such claim was ever preferred is significant of the difference between the newer generation of Englishmen at the Antipodes, and the older and more mixed race in the Americas. In the United States the millionaire is a cult. The Vanderbilts, and the Astros, and the Mackays, and the Goulds are less accidents than institutions. Their private history is followed with a keenness of investigation unknown in other lands. No American citizen so poor that he cannot glibly reel off in breathless haste the exact details of the quarrel and the reconciliation of the Vanderbilt, junior and senior; or the real reasons why the rising hcope (sic) of the Gould's sacrifices a million on account of his union with a popular, if not too Juvenile, actress. But in Australia the case is different. We have had but one millionaire of what may be termed the first rank. And now that we come to sum up his biography, we realise that the history of a man, absolutely unique among his fellows, rests on mere oral tradition— is, in short, a mere record of hearsay, handed down from lip to lip, that the one or two people who were ever admitted into his inner confidence, if any such exist, are

---

[1] James was 79 at death: Zita Denholm, T.Y.S.O.N. Triple D Books Wagga Wagga 2000, pp.21, 30.

[2] James was born in 1819: Zita Denholm, T.Y.S.O.N. Triple D Books Wagga Wagga 2000, pp.21, 30.

dumb on the subject. and that, in short, while everybody appears to know of him, no one seems to have known him.

There remains of the earlier days of James Tyson not one word. He stalks out first into the light of day a tall, lanky youth, between six feet one and two in height, then, as most cornstalks, an overseer at L30 (£30) a year and his fill of ram-stag mutton and heavy damper, but yet silent, taciturn, and full of purpose. In 1846 so rapid were the changes of those days, we find him a full-fledged squatter, partnered with his brother, William, between the Lachlan and the Murrumbidgee, and five years later, the goldfields having meantime broken out, he grasps the money-making advantages of selling butchers' meat over the hard work of digging for gold, and so lays the foundation of his fortunes. He was then twenty-eight[3]. Had he ever had a single weakness of youth? Had he ever yielded to one of those generous impulses that boyhood classes as virtues, and mature age as follies? No one can tell. Even then he knew to the full how to guard the impenetrable riddle of the Sphinx. And he died as he lived—silent, in secrecy, no man knowing.

It is natural, when so little facts, are known, that there should be two theories of his life work. These are excellently exemplified in the morning dallies. According to the "Argus," the deceased was a man of the most kindly and generous instincts. The "Age" remarks upon his parsimony. "He was a bachelor," says our contemporary, "most economical in his personal expenditure, and a total abstainer from wine, spirits, and tobacco. In 1892, in a time of great financial strain for the colony, he took up L250,000 (£250,000) of Treasury Bills, in order to assist the Queensland Government." There may appear in these sentences to be something of a contradiction. And yet

---

[3] Thirty-two: Zita Denholm, T.Y.S.O.N. Triple D Books Wagga Wagga 2000, pp.21, 30.

the story of his life was never read better. It is claimed by one school that he was a misogynist and a miser. Another asserts, with equal force, that his private charities were large and secretly bestowed; that, veritably, his right hand knew not what his left hand did. Which is the true theory? Who knows?

We have heard it said of him— and the story is so generally known as to need no excuse for mentioning— that he once walked from Adelaide to Sydney, starting with half-a-crown in his pocket, and finishing his journey with the same amount of capital. From this fact he got the familiar name of the "grass-eater." The probability is that he "worked his passage," for he was never a man afraid of work. No "Jackaroo" on any of his numerous stations began earlier or left off later. And the probability is that in the early days his peculiar method of dressing laid the foundation of the many yarns about him which are now classics round all camp-fires, and among all the itinerant tribes of swagmen from the Leeuwin to the Gulf, He has sold himself that in his early manhood "they didn't worry about clothes. A sack with three holes, one for the head, and two for the arms, was good enough for him." In later years, and when the necessities of his many interests drew him into the closer communion with towns, he made new sumptuary laws for himself, and was always well, if plainly, dressed. But to the last his private expenses were strictly limited, and his indulgences absolutely nil. He stood apart from most human relaxations. The pleasure of lifting one million on to the top of another, as one crowns the successful king in the game of draughts, was enough for him. It may be that this is the most supreme joy of all, wonderful, passing the love of woman. We who have never trifled with seven figures cannot tell.

To the last he was a simple man. In his hardier manhood he never stopped at a hotel if he could camp out and cook his simple damper and billy of "post and rail tea." Was this meanness, as some say or a poetic love of the bush? A New South Wales bishop once appealed to him for a subscription for the most excellent ends and got a L5 (£5) cheque. The reverend diocesan promptly replied that he expected more, and was gratified to receive a polite note, saying that a mistake had been made, and asking for the return of the cheque. Of course, the cheque was promptly returned to the donor, but the right reverend Anglican, if he be not already translated to a higher diocese, is still waiting for the four-figured cheque which he calculated on in return.

There are many other anecdotes, all of which will doubtless leak out in time, but which for the present may be deemed out of place. The mystery of the man's life remains behind him. How much does be leave? Curiously enough, no one dares but to speak of the sum in millions. It may be four or five, or three to four, or two to three. Anything under two would be treated as a breach of faith. And does he leave this stupendous total to his numerous nephews, or to some public end? There are men in the city now who whisper that he had often said that, as he had made his money out of the Queen, he would leave it to the Queen. Others translate the words that: "He had made his money out of the country; he would leave it to the country." And, in the end, as in the beginning of his life, we say once more. "Who knows?"

Who knows! The Herald was confused about James Tyson but not E J Gardiner:

The Argus, Melbourne, Friday, 9 December, 1898 P 5.

TO THE EDITOR OF THE ARGUS

Sir,--Kindly allow me to supplement your accurate report of the late Mr. James Tyson. From a pleasant conversation I had with him in Queensland some years back. I gleamed several interesting details as to his phenomenal success in life—a man having great force of will and determination of character, a strange and peculiar individuality; a man with a big heart, full of sympathy and human kindness, although he was supposed, unjustly by many to be near and penurious. Not so he seldom made his benefactions known, always avoiding publicity in regard to this matter. His life was a peculiar study.

Never actually trustful of others, he had but few confidants; as he remarked to me, he did not require them. He had an iron frame, a physique and stature gaining admiration. The late isolated man has gone to his rest. His desire was not to heap up wealth. A significant remark he made to himself was this, "I have a star in the distance to which I must attain," then probably the true character of James Tyson will be known. I inferred that in the future he intended to make large contributions to educational and other institutions.

He would not tell me the nature or direction of the star, but I am convinced, from our discursive conversation, should he have left a will, his immense wealth will not be all absorbed by his relatives.

His first success in life, as he told me, is correctly given by you—10s. per week, long hours, hardships; no eight hours in those days. With the first few pounds he saved he traded with station hands in small articles, such as knives, tobacco, and numerous trifling articles. This became the foundation of his great wealth, gained by close application, industry, strict rectitude, and honesty—and not a forward man in public or private life. His peculiar personality was known and respected by all inhabitants in Queensland, and many in other Colonies. G.A. Sala, in his lecture, spoke of men I have met, but he never met James Tyson.

Yours, &c.,

E. J. Gardiner.   Wellington Street, St Kilda, Dec. 7.

It is interesting to consider who the real James Tyson was and if E J Gardiner's comments are representative of opinions at that time. However, it is of concern that he mentioned "peculiar personality" three times in such a short letter without saying what is peculiar about him. This adds to what I want to find out about James.

The following was typical of many provincial newspapers written by reporters who had never met him:

Ovens and Murray Advertiser, Saturday, 27 Apr 1889 P 11.

AN AUSTRALIAN MISER.

The "prosperous Queensland squatter, whose landed possessions are estimated at five million," mentioned in the late Lady Brassey's posthumous volume, is Mr James Tyson, at once the wealthiest and the stingiest man in Australia. Almost every second individual yon meet in the colonies has an anecdote to tell of the " Hungry Tyson," to give his unenviable sobriquet. An Australian prelate who was in London recently, attending the Lambeth Conference, narrated how, in a voyage along the Australian coasts, Tyson, who was a fellow-passenger, slept on deck and munched sandwiches the whole of the way, a berth and meal from the ship's resources not being in accordance with his ideas of economy. Hitherto, Tyson has been but little known or heard of in England, and Lady Brassey is the first to prominently call attention to the existence of a colonist who has become wealthier than either the Duke of Westminster or the Duke of Devonshire, and who owns more land than the two of them put together. Lady Brassey localises him in Queensland, but he has enormous landed interests in Victoria and New South Wales as well. The five-times millionaire is now about seventy years of age, unmarried, and goes about like a day-laborer, in the commonest of clothes. No one has the remotest idea into whose hands his vast wealth and estate will fall upon his decease. [This modern Croesus is just now in the Upper Murray district. The wife of a Bonegilla publican managed to sell two art-union tickets to him the other day.]

Parameters

This work has three parts. Part 1 to page 181, considers the rumours that portrayed James as a miserly, bitter, machinating, misogynist and a "pretty stunted specimen of humanity". Part 2 to page 227, looks at specific rumours, as listed, with a chapter for each one. Part 3 is from page 228 and has important observations.

I kept every article as it was written except for items unrelated to James Tyson, which I did not include. This ensures the message and style of the author and period is retained; however, it also means there are duplications. The story of James Tyson is repeated many times in the newspapers; therefore, if a number of articles are quoted for their difference, there will be some duplications because they have the same backstory. This is a small price to pay rather for the risk of editing or changing the message intended by the author of the article.

Part 1 is covered over six chapters; each one is affected by significant events which would be expected to influence community views about James. Chapters 1 to 3 has articles that were published before James died so they were open to his scrutiny. Regarding the interview outlined in Chapter 2, James made comment to the editor, Charles Buzacott, who testified in the Domicile Case James said to him, " It's all right, mister; I believe it's done good; I believe it's done good," Signifying he had read it and was obviously happy with what was written.

CHAPTER 1

Articles to the first and only formal interview of James Tyson

Chapter 1 traces the population coming to grips with this strange man. A man who made money out of nothing and was able to keep it. For instance, in this chapter, you will see a story how he backed his judgement and bought a 20,000 mob of flood-bound sheep on the chance he could get them out of their terrible plight. He did. Then, took them to the gold fields to sell for a big profit.

Starting in 1855, when James and brothers sold their gold fields operation, and continues until 1893, before the only formal interview was reported. This chapter is by far the biggest in time spanned and biggest in the activity of James Tyson but the least in writings by how others see him. He invested in sheep, cattle and sugar. He went from one property to dozens spread over three colonies; and, he usually visited each annually just travelling with horse power. James went from struggling grazier to the richest person in Australia and a world-wide curiosity.

James would have been 36 years old when the following article was written. This appeared to be the first time James appeared in print other than in a paid advertisement. He purchased the Deniliquin properties in November 1855 so it did not take long for him to be involved in local activities.

Sydney Morning Herald, **Wednesday, 23 Jan 1856, P 5.**

EDWARD RIVER DISTRICT. Our CLERGYMAN.

The contributions towards a resident minister for the Edward River have now about reached the necessary sum. Mr. James Tyson has kindly offered the free use of a neat cottage attached to the chief station of the late Royal Bank property, at North Deniliquin, as a residence for the expected pastor, whose presence will be welcomed by all good men in the district.

The below surprising news item first appeared in Southern Courier (a Deniliquin newspaper at that time) and was copied in many other newspapers.

Mount Alexander Mail, **Friday, 19 Oct 1860, P 3.**

WATER SUPPLY ON THE PLAINS.

Mr James Tyson, having completed a very large reservoir at the ten-mile station on his Upper Deniliquin run, invited his neighbours to meet him there to view it on Tuesday last. Owing to the wet weather and the heavy state of the roads, the gathering was not so numerous as there is reason to suppose it would otherwise have been, considering the paramount importance of the question of securing permanent water on our back plains. Two settlers from the Billabong put in an appearance: the Murray and the Tuppal also contributed their quota. All present declared themselves highly delighted at the noble sheet of water presented to their view, and heartily approved of the plan on which it was, constructed. It may be taken as a proof of what our squatters would do towards improving their runs had they security of tenure; and we may remark, 'en passant,' that by this and other improvements Mr Tyson has made, his run can carry four times as much stock as it did when it was under the Company's management.

The tank measures 130 feet long by 70 ft. wide, and contains 16 feet of water, clear as crystal. It is lined throughout with pine trees, and is sloped at each end to allow sheep to drink out of it. The water is collected from swamps in the vicinity, by drains, one of which is a mile and a quarter long. These drains were made by the plough and spade, are about three furrows wide, and are carefully finished off, so as not to allow the water to wash any dirt away with it into the tank. The immense quantity of earth thrown out in making this reservoir looks at a distance like a railway embankment. All agree in pronouncing the tank a great success, and in landing the care, skill, and forethought manifested in every part of its design and construction. It would be well worth anyone's while who takes an interest in the modus operandi of tank making, to ride out to see this first-class specimen. They would confess themselves amply repaid for the trouble, even although they should not find, as Mr Tyson's guests did on this occasion, a champagne luncheon awaiting them, — Southern Courier

I'm sure many people who knew James later in his life would be amazed to hear that he hosted a champagne luncheon! but, maybe this was the real James Tyson.

Consider the next article – Separation Committee – There was a move for the south western part of the Colony to secede from the colony. This relates to the southern-western part of the present state of New South Wales. Some of the members of the movement wanted to align with the now Victoria or make it an entirely new colony or state. It shows that James was very keen to be involved in local issues by giving one thousand pounds which would have been close to a million dollars today.

Maitland Mercury, **Thursday, 14 Mar 1861, P 4.**

SEPARATION COMMITTEE

The Deniliquin Separation Committee has appointed a paid secretary to correspond with the committees elsewhere formed, and with individuals favourable to the object. One squatter Mr. James Tyson, has promised £1000 as his subscription, it was stated, "if the agitation was carried on as it ought to be." The Pastoral Times understand another meeting is to be called, to initiate a Separation League. The Southern Courier says much disappointment is felt by the squatters about Deniliquin at the manner in which the agitation has so far been conducted.

The Goulburn Herald and Chronicle, **Saturday, 1 Oct 1864, P 3**

LOCAL AND PRO VINIAL.

Magistrates.-His Excellency the Governor has been pleased to appoint the undermentioned gentlemen to be magistrates for the colony:-- George Jackson Frankland, Mowbray, Paterson; Abraham Nivison, Walcha; George Barnet Townshend, Trevallyn; and James Tyson, Esquires.

Argus, Tuesday, 16 June 1868, P 7.

WATER CHANNELS

Mr. James Tyson, remarks the Riverine Herald, "after having done wonders on the Lachlan in cutting canals, erecting dams, &c, has extended his operations to the Warrego and Upper Darling districts - having purchased several stations in those distant quarters. Mr. Tyson is a practical squatter, and having thereby, by self-denial, industry, and an intelligent direction of large means accumulated a large fortune, spends much of his surplus cash in improving the Crown lands, at the same time profiting thereby. Mr. Tyson seizes on a barren country, and it is immediately covered with flocks and herds."

Argus, Melbourne, Thursday, 19 February 1885, P 6.

THE CAMPAIGN IN THE SOUDAN.

Today the following munificent offer was received by telegram by the acting Colonial Secretary from Mr James Tyson, of Felton Station, Cambodia River -"To the Hon. W. B. Dalley,-I will contribute £2,000 per annum towards the expenses of the Australian contingent in the Soudan, for two years if required I hope that nothing but really strong and reliable men will be sent that will make their mark and do credit to Australia generally. No distinction should be made between the various contingents from the Australian colonies, which might create bad blood between them -JAMES TYSON "

The person who planned and financed Practical Domestication was James Tyson

Toowoomba Chronicle, **Tuesday, 30 Aug 1887, P 3.**

PRACTICAL DOMESTICATION.
The Result of the Competitive Examination.

In the month of May last a valued subscriber to this journal advertised in our business columns the fact that he would give two prizes, of £50 and £25 respectively, for a competition in practical domestication open to the young women, of the Darling Downs district. His object in offering these prizes was, that, as a keen observer of the tendencies of the present age, he noticed the want of that training which alone prepares young women for their true sphere of good house wives and mothers. He thought that by offering prizes he might, by drawing out the qualifications of the competitors, arouse others to a sense of their responsibilities as the mothers of coming generations and of those in whom is mainly centred the destinies of a great nation. The gentleman in question placed £100 in the Royal Bank of Queensland to the credit of Mr. G. P. M. Murray and the Hon. W. H. Groom, as the prize fund. The rules of the competition were as follows:-- (1.) Age from 15 to 21 years. (2.) Figure, strong, well-grown, healthy, with good carriage. (3.) To be able to read English well aloud, to write a decent letter according to grammatical rules, and perform on some musical instrument or sing moderately well. (4.) To be able to cut and make their own dresses and underclothing; dresses to be neat, no artificial means (such as pads, steel hoops, high-heeled boots, &c., &c.), to be allowed. All intending competitors were to send in their names by the 30th of June. Nine young ladies entered, and the competition took place on Friday and Saturday last. Eight young ladies attended for the examination which was severe in all the subjects. Dress material and calico were provided from one of the business establishments in town, and the competitors were required to cut out dresses, make them, and prepare other articles of apparel, pertaining to a lady's wardrobe. In the food preparation competitions, the workroom was changed for the kitchen, and the competitors were requested to make beef tea and sago according to their conception of these medicaments for a sick room. In the general cookery each competitor was asked to write a paper on their ideas of the viands and their preparation which should make a good plain dinner for 12 persons. Added to this theoretical test was the practical one of making a beefsteak pie. It is said by the faculty and by hospital nurses that there is an art in making beef tea; if so, what can be said of making a beefsteak pie? This test was submitted and it was a good one. Going from this, another step was made from the kitchen to the drawing-room. The competitors were asked to write a letter to a parent, brother, sister, or friend. A reading aloud test was also included, the piece chosen for each competitor being a selection from Dugald Stewart's "Noble Thoughts in Noble Language." Music, too, was in the list, and all played pianoforte solos and sang both from sight and memory. The young ladies were also requested to enter and retire from a drawing-room in accordance with their ideas of deportment. This was the extent of the examination, the judging in which lasted two days. A good system of judging was adopted, one which could not be disputed. A standard was fixed and the judges, both ladies and gentlemen, worked to it. The first prize taken reached this standard with one slight exception; the second prize-taker was but a short distance behind; and a third prize was awarded from the balance of the £100, any part of which the donor refused to have returned, which remained after the necessary material for the competition had been purchased. The prize-takers were as follows:

Miss A. D. Broadfoot (Toowoomba) £50 0 0

Miss Elizabeth Sims (Toowoomba) £25 0 0

Miss Ann J. Sanders (Warwick) £7 10 0

The whole of the competitors acquit ted themselves so satisfactorily as to realise the expectations of the donor. The result, indeed, has been so satisfactory that there is every probability that the competition will be an annual event, the funds being supplied by the same gentleman. It may be added that the work performed was excellent, and the first prize was unanimously awarded. The winner showed high capabilities in all the branches of practical domestication in which she was tested, and her letter-writing was a marvel of neatness and excellence, while in reading, playing, singing, and cookery she was surpassingly clever.

Sydney Morning Herald, **Tuesday, 15 Mar 1892, P 6.**

The interview between Mr. James Tyson and the Queensland Treasurer is one of a kind which it would be well to see more frequently illustrated in the public life of these colonies. The case is one of such interest that it will be by no means space wasted if we here repeat the telegram on the subject we printed on Saturday last. It is as follows:

"It is understood that Mr. James Tyson waited upon the Colonial Treasurer, and said he was desirous of assisting the colony if Sir Thomas M'Ilwraith could say in what way he could do be. Sir Thomas suggested that if he would purchase some of the Queensland Treasury bills his action would be greatly appreciated alike by the colony and the Government. Mr. Tyson inquired the price, and was informed that it would be such as to produce 4½ per cent, interest to the buyer. Mr. Tyson at once intimated his willingness to take £200,000 worth of bills, stating that he would pay £100,000 at once and, the balance on demand."

Now, it would be an over-estimate were we to claim for

this act the character of rare or extraordinary patriotism. Rare in one sense the incident is, no doubt, since it would be necessary to search wide and far in our colonial annals to find a similar instance of a man of large wealth coming in a like way to the assistance of his colony in a time of difficulty. But when we remember how much in so many ways society does for the possessors of great wealth-wealth which is largely augmented, even without the labour of the owner, as a result of the progress and security and enhancement of value caused by the work and sacrifices of society as a whole-it is not a satisfactory reflection that this wise, patriotic, and timely act of Mr. Tyson should be as rare and as solitary as we must recognise it to be. No doubt Mr. Tyson gets for his money a rate of interest and a security which cannot be regarded as unsatisfactory. But the quality of this proceeding is not exhausted when it is viewed on strictly business considerations. The fact remains that in a time of difficulty, depression, and distrust, Mr. Tyson asked to be shown how he could assist the colony, that practical Sir T. M'Ilwraith promptly suggested that he should take up some of the public securities open for investment; and that the suggestion was adopted to an amount which will at the present emergency be welcome to an embarrassed Treasurer.

The case is gratifying from the patriotic motive which animated the act, and if its result should be to induce imitation, and to lead other men of great wealth to come to the relief of their respective Treasuries in a time of difficulty, and thus to do a service to the country which has done so much for them, the effect would be beneficial in more ways than one. It would more closely than ever identify the interest of general prosperity and the interest of the wealthy class, and by associating wealth with patriotism and with the recognition of the principle that as noblesse oblige, so also does the possession of great wealth impose great responsibilities, it would add to the security of property, which is never in such jeopardy as when that security is based on some ground different from that of the interest of the general community.

The Bulletin Vol. 12, No. 635, (16 Apr 1892), P 21.
A Much-Misunderstood Millionaire.
[FOR THE BULLETIN]

A tall man, solitary in his ways, always dressed in grey tweed and invariably wearing a soft felt hat. A healthy, ruddy complexion, iron grey beard and moustache, with extremely bright blue eyes. These last are a noticeable feature and compel your attention from their peculiar clearness.

When you pass this, individual, who is almost certain to be alone, for millionaires are a lonely race, you have passed the richest man in Australia, James Tyson, who has just come into special prominence by the substantial financial help he has accorded the 'Queensland. Government.

Ninety-nine out of every hundred people pass "Jimmy" Tyson without noticing him, for although everybody has heard of him, personally he is little known. His dress does not betray him, either by its loudly-proclaimed magnificence or by any assumption of careless neglect. One hears it often said that the mammoth millionaire habitually clothes himself in slops or moleskin and munches a roll and a polony for his lunch. Nothing of the sort. His clothes are tailor-made and well-fitting, though very sober in colour, his only singularity may be said to be his affection for a soft felt hat as head-gear. He has his club and lunches there like other men.

Tyson belongs to a generation now giving place to somewhat effete successors. He is one of the early race of "natives," who first caused it to be generally supposed that all Australians are tall, gaunt, sun-burnt, and taciturn—-a race who did good work in their day, and many of whom are deservedly enjoying the fruits of their early struggles with the wilderness. It is true that many who also struggled hard and honestly are now in the Benevolent Asylums.

James Tyson and his brothers were born at Menangle in N.S. Wales, and took an active share in the pioneer settlement of Australia, Earth-hunger is the family failing. One brother, Peter, kept himself for years an outwardly poor man, through his eagerness in buying up land with every shilling he could muster. Of the subject of our sketch more yarns, fabulous arid otherwise, have been told than of any other private individual in Australia. Most of them are chestnuts applied indiscriminately to any rich man suspected of nearness, and others are palpable lies.

We have no wish to pourtray (sic.) king-monopolist Tyson in the light of a saint, ready for a halo and enshrinement in a stained-glass window; but so many of the stories concerning him are greatly misleading as to his true character, that a few veritable facts about the chief millionaire of Australia may prove entertaining. James Tyson loves the bush, is essentially a bushman, and is always more conversational on subjects relating to country affairs. The only time the writer of this article ever had a lengthy talk with him was in Brisbane, just after returning from a prolonged trip out West. The conversation turned upon hardship and short commons in new country, and Tyson grew quite eloquent on the various animals and vermin that could be used to support life after the ordinary rations had failed. He expressed a decided disapproval of possums; bandicoot was what he swore by.

Tyson's objection to being imposed upon, and his unostentatious ways have gained him an undeserved reputation for stinginess, but too many well-known instances of his generosity and gratitude (which latter examples almost invariably take the form of liberal aid to men in difficulties whom he knows to be resolute and honest) are extant for this to be the case. But he is obstinately determined not to be "had" as a certain church dignitary in Queensland found to his cost. The gentleman in question was pushing a charity forward, and Tyson sent him a cheque for £250. The reverend sent back a polite note thanking him for the donation, but intimating that he had hoped for a larger sum from a man of Mr. Tyson's known wealth. Tyson, in reply, asked for the cheque back. It was readily sent, the clergyman smilingly expecting a larger one. It never came. Tyson was disgusted at such barefaced begging, and now that parson never omits an opportunity of assuring his hearers that a rich man cannot get into heaven.

It is related of him that he always travels by steamer under an assumed name, after the manner of Jonah Perkins, in order to avoid any possible application for a loan; but it is not the case, any more than the hackneyed story that when traveling in the bush he carries a few cobs of corn in his saddle-pouch, and camps out and boils them in his quart-pot in preference to putting up at a station or a roadside inn.

Tyson's peculiar habits, unconventional, solitary and sparing of speech as he is—by the way, he invariably calls everybody "Mister," no matter how long he has known him—have been the cause of much of this misconception. He is supposed to be a woman-hater, but people who lived out on the Warrego during the early days tell a different story, and though our multi-millionaire has never been guilty of the weakness of matrimony and absolutely shudders at the restraints of fashionable society he will tell you, if he knows you well enough, that he yields to no one in his admiration (from a respectable distance, certainly) of the cultured and healthy and cheerful woman who is, or is fitted to become, a prudent, helpful and virtuous house-wife. It must be said, in extenuation of, James Tyson's bachelorhood, that he is the only one of his family against whom there is anything like the shadow of a prima-facie charge of misogyny. No jury would convict his brother Peter on such an accusation. Nor his now illustrious nephew, James.

Coming down in the steamer from Brisbane to Sydney, B-------, a well-known and genial Queensland land squatter, shared the cabin with Tyson. As a rule, the confined space of a cabin necessitates one man dressing at a time; B------- was dressing, and Tyson lying, meditating in his bunk.

"B," he said, at last, "you are not half the man you were. You are getting like my nephew, J-—, too fat; you eat too much, and drink too much. Why don't you live naturally?"

"Do you believe in living naturally?" said B------.

"Certainly, I do."

"Then get married as soon as you can," replied B, as he left the cabin. Tyson's reflections on the subject have not been recorded.

The same man, who was a neighbour of Tyson's, relates that, being in want of a sheep-drover, an individual, whom we may call Blanks, applied for the billet, and mentioned Tyson as a former employer. On being applied to, the masterful James gave the following reference: —

"Blanks! yes. He took some sheep down for me, and did very well with them. I was giving him thirty shillings a week. I raised him to two pounds and gave him another lot. He did all right, but not quite so well as the first time, and I noticed that he commenced to wear a collar. However, he got another lot, and a rise of ten shillings. This time he did not do so well as before, and he had started a necktie with a pin in it. Well, Mister, to make a long story short, he got another job and another rise, and then I saw him buckling on a pair of long-hooked spurs, and he had about four pouches on his belt, but he didn't do anything like so well. No, Mister.' Blanks is a first-class drover at thirty shillings a week, but at two or three pounds he's no good. Success spoils him."

In the sixties, when a race meeting was being held in Hay, Tyson passed the place. He camped out, and his horses wandered away. Next morning B------ met him with his saddle and swag, on his shoulder, going down the main street. "Why don't you stop and wait for your horses?" he asked. "No," returned the stubborn James, "I never turnback." He walked on eighteen miles before the black boy overtook him with the lost horses

An instance of alleged nearness has often been told regarding one of his western stations, whereon, it is asserted, he refused to allow any milkers to be broken in, on the ground that every pound of butter made and every quart of milk drunk robbed the calves of flesh. The truth and origin of the yarn being that, visiting the place during a severe drought, he found a number of poor milkers and half-starved calves kept in at a time when the paddocks were destitute of grass, and naturally ordered them to be turned out.

All the money he has made in pastoral matters has been acquired through good judgment and shrewd management. Long ago, when fat cattle brought a decent price in Melbourne, he had secured good fattening paddocks in one of the best districts of Victoria, and regularly overlanded his store bullocks and had some always ready to secure top prices in the Melbourne market.

His method, and love of care, pursue him through all the avocations of his life, as witness his careful remembrance of the fact that he owed the barmaid of a leading Sydney hotel ninepence. After an absence of nearly twelve months, he duly called in to pay it. Not to be beaten, the girl handed him a pair of socks which she asserted he had left behind after his visit, and which she had had washed and darned. Naturally Tyson took them with unmoved gravity or he would not have been Tyson.

One day when returning home on one of his Queensland stations he overtook a couple of swagmen, men, who asked him the distance to the homestead, which he told them. "It belongs to that hungry ------, Tyson, doesn't it?" said one. "Any chance of a feed there?" "You'll get a night's shelter," said Tyson, riding on, when one called out, "Got any matches or tobacco on you?" Tyson had both, although he does not smoke (all the same, it is his delight to cut up tobacco, that which he sniffs lovingly, as if he did indulge in the weed). He gave the men a fig of tobacco and some matches. At sundown the swagmen turned up and the storekeeper gave them the usual rations. Just as he had finished Tyson came in to the store and remarked, "These men owe me 8d.-—6d. for tobacco, and 2d. for matches."

"Who the carnation are you?"

"I'm that hungry -----, Tyson," he replied. "If you had been civil, you might have had the tobacco and welcome, but now you can pay for it." And they did.

Tyson's first step towards fortune was made, he relates, far back in the Fifties. He was taking stock over to Victoria and found himself stuck up by flood and bog. Near him a sheep-drover was camped under similarly wretched circumstances, and after some talk Tyson bought up his sheep at 6s. per head. The purchase completed, he set about finding means of transit across the flooded river and obtaining a suitable-boat succeeded in crossing both his own cattle and the sheep without loss. He made for Bendigo, then in its glory, where he sold the sheep (16,000) at 26s. per head, and the cattle at 60s. per cwt. Since then, by judicious investments, and a ready grasp of every situation where there was an opportunity for a good bargain, he has made the colossal fortune with which he is generally credited. His capital is variously estimated at anything between four and ten million, and it is stated on good authority that one year's operations in Queensland and New South Wales netted him at least £425,000. This is a tidy income, and it is not to be wondered at that his fame has travelled beyond Australian shores, and that he was once made the. subject of an article in a London paper, wherein a lady trotted him out as a matrimonial prize worth trying for.

With the family eye on land, Tyson has been interested in syndicates for land-grant railways in Queensland on several occasions, but, as the projected lines have never come off, his estate is still lacking the ample dimensions he would seem to desire. What will become of his wealth is a question. He has no relative who has any great claim on him, and he is a hale, hardy man of 60, with every faculty unimpaired, and as keenly bent on making money as ever, so that no one can estimate what his fortune may be when his time comes at last. He has been heard to lament on several occasions that his tastes did not lie in some direction outside of stock and finance. Even horse-racing would be welcome if he could only bring himself to take a thorough interest in it. " But the bent of his disposition does not tend that way and he has to content himself with turning over a few paltry hundreds of thousands every year. A millionaire who does not drink, does not smoke, is impervious to the most fascinating wiles of fascinating women; and yet is not a sordid miser of the Daniel Dander type, is something unique, and Australia may take credit for having produced the first specimen of the kind. He has not. even built himself a "lofty pleasure house" after the fashion of Mount Morganaires or Broken "Hillionaires, one of whom has lately had designed a dwelling-house which is a conundrum. The puzzled spectator on seeing this design says, "O yes. A new town-hall." "A new theatre!" "A lunatic asylum!" until at last he is told that it is a simple dwelling-house, and is struck speechless. Tyson is too much of a genuine bushman to feel any pleasure in such empty magnificence. He dislikes society. Government House has never had any attractions for him. Even the "genial" Carington could not get the reserved and silent Jimmy to put his legs underneath his mahogany, although he tried personal persuasion and Lady C. wrote an invitation in her own fair hand.

As already remarked, Tyson is one of the old race of natives who grew up to self-reliance and hard work. Their paths were not made easy for them by a rum-selling progenitor, or a successful land-booming parent, but they learnt their bushcraft from Mother Nature, and have been her children ever since. To this may be ascribed the fact that in spite of Tyson's business shrewdness and natural disposition towards making money, he retains his love of simplicity, of fresh, breezy nature, and the acquirement of knowledge of all sorts. An indefatigable reader, he has probed problems with which no outsider would ever suspect him of meddling.

Once, during his early days, he was driving a horse-dray, and at the foot of a hill the horses, which were pretty well done, stuck him up. A teamster was passing, and from him Tyson asked the usual favour of a pull up the pinch. The driver, seeing he was in a fix, asked a price which Tyson justly considered extortionate, so he declined to give it. He unloaded the dray singlehanded, handed, and by dint of hard work and cunning zigzagging, got it to the top of the hill; then he got the load up piece-by-piece, and finally led the knocked-up horses up and proceeded triumphantly down the other side, having successfully resisted an attempt at imposition.

One evening a well-known bush-missionary (self-dubbed) came to one of Tyson's stations when he happened to be at home. The individual sent in his name and was received with the usual bush hospitality. He improved the occasion after dinner and extracted from Tyson a half-promise that he would contribute towards his little mission. Unfortunately, there was some rum on the place, and, at the invitation of the manager, the "servant of the Lord" took two or three second-mates" nips that put him to sleep in one of the verandah chairs. Next morning, he reminded Tyson of his application for a subscription, but that individual grimly remarked that as it was a dry country, the kindest thing he could do was to give the shepherd the perpetual free run of the tank —the water-tank, not the rum-keg.

We can scarcely say that it would be well for the country if there were many Tysons in it, but there is no doubt that the same sagacity and earnestness, the same splendid generalship and iron determination which have served him so well at money-making, would have brought him to the front rank in many other pursuits. His good deeds are little known, for he does not advertise them. Now and again, one hears of them by accident, but everybody is ready to repeat some alleged anecdote of his dog-mean close-fistedness.

CHAPTER 2

The only formal interview of James Tyson on 8th December 1893

We have to trust that the interviewer faithfully recorded the answers. Of course, at the time of the interview there were no recording devices but Pitman short-hand was in use. As I read it, I did get the feeling that the tone of answers was a little different to the tone of the questions. This suggests that the answers were taken from a form of short-hand rather than memory. At the end of this chapter, I give comments from James to show that he read the finished text and verifies it.

Brisbane Courier, **Friday, 8 Dec 1893, P 5.**

OUR MILLIONAIRE

Erect despite the burden of seventy years, as a Red Indian Chief, standing over 6ft. in his stockings, neither spare nor stout, well-built and firmly knit perhaps slightly angular with unshaven silvery beard and locks thinning by the hands of time; with keen searching eyes, and apparently somewhat embarrassed by the shyness of a bushman unaccustomed to contact with civilisation – my subject presents himself to the reader as a man well preserved, active, self-contained, and of perfect physique. "Born in New South Wales, on 10th April, 1823[4], I never took a dose of medicine, never had a day's illness, never lost an hour's time by sickness in the whole course of my life." The result of frugal living, high thinking, hard work and of a country life in the most healthy climate under the Sun.

Our millionaire is not a popular man by thousands who do not know him, or who picture him as the typical monster of plutocracy, he will be condemned without trial, and figuratively hung, drawn, and quartered as a culprit unfit to live. As an Australian Jay Gould he may be exhibited as the meet subject of popular execration. But, as a matter of fact, he is not a monster, is not in any sense a bad man, bears no resemblance to the big American gambling speculator, and can be accused of no crime except that of having accumulated the largest fortune in Australia. Beginning life as a working overseer on a station at £20 a year – remuneration which a labour organiser would denounce as insufficient for a b*********w – my subject has lived a life of hard labour and self-denial, and does not possess a shilling which he has not honestly earned.

Our millionaire is thrifty but not miserly, acquisitive but not niggardly, enterprising but the reverse of speculative. For him the stock exchange has no attractions, the gold mine has no temptations. A great landowner and grazier, he sticks rigidly to the business of his life – the reclamation of the wilderness, the multiplication of sheep and cattle, the enlargement of the pastoral capabilities of the great Australian interior. If he possesses enormous bank deposits, it is because he prefers low rate of interest to the worry and trouble of money-lending in the ordinary manner. But his money is in no sense locked up unproductively, or "lying idle" as the popular conception goes. It is helping the industries of the country, and earning interest

---

[4] This date is incorrect. James was born on 8 April, 1819. In *Australian* *Dictionary of Dates and Men of Our Time* by Henniker Heaton lists his birth as 11 April, 1823; I deal with this later.

every day. Our millionaire, though he does not carry on any great industry in person except that of squatting, thus conducts many industries by deputy, and the state of interest he receives or covenants to receive, is exceedingly moderate. He spends almost nothing on himself, but all his money goes into circulation as freely as though he scattered it with a lavish hand. His conception of the duty of man is that he should direct his labour entirely to useful employments, and his money into solely reproductive channels. Money poured down a man's throat in costly and injurious liquors, or wasted in merely ostentations luxuries, might almost as well be thrown into the sea. So spent it is, except such percentage as may be secured by the local trader, simply capital annihilated. Whereas the money well vested by the millionaire accumulates and multiplies by its employment in useful industries and in the development of latent resources.

Our capitalist detests idleness and extravagance as Nature abhors a vacuum. He holds that muscle, brain and money were made for use, and not for abuse, or disuse. He has no money upon the loafer, or the helpless parasite who "multiplies children, giggles at girls, and lives upon other men's industry." A self-contained, solitary, strong, patient, irrepressible man is our hero, born to grapple with Nature in her most stubborn moods – to circumvent droughts, combat floods, subjugate the wilderness, transform the desert into a fruitful field, and the arid bush into a fragrant garden.

Our millionaire is a man of foresight, of mental power, of unbounded faith in the resources and the future of his native country. A man who believes that Central Australia's one wants in men – of physique, industry, grit. Of men who like himself make life a winning game, and will pursue it in the face of angels and devils, and to the very margin of the Eternities. With him life is work and labour the primary if not the whole duty of men. Taunted with the folly of piling up wealth for an unknown heir, our millionaire is reported to have said, "If my successor derives half the pleasure from spending that I do in accumulating, he will be happy." Not one of us, his critics, would take life as he takes it, or would spend our money as he spends it. We would devote it to charitable objects, to assisting deserving young men, to promoting new industries, or to keeping up a luxurious establishment. But are we quite sure that we would in these ways spend our money better than he spends it? The man who would spend money well must first learn to earn it and save it. Our millionaire has learnt to do both; he thinks he spends it well. We *might* do better; but then we are not millionaires, and with other areas of obligations never shall be.

Our millionaire has, however, a living sense of duty to his country. Naturally of a retiring disposition, he refused all political honours until seventy summers had overcome his reluctance and matured his judgement. But on attaining the ordinary human limit of years he was induced to become a member of our Legislature, and as such he recognises that his influence ought to be exerted for the benefit of his country, even if to do so he must lift the veil of privacy with which he has hitherto jealously invested his life and works. His place of residence is Felton, Darling Downs, but he is nearly always on the move, and passes frequently through Queensland, New South Wales, and Victoria, in his travels. He possesses a more extensive knowledge of the interior of Australia, of its climatic and other characteristics, and of its dormant resources than any other man living. He has unbounded confidence in its future; his opinions are entitled to the weight of conclusions painfully elaborated in the rough school of experience; and he has at length been convinced by entreaty and argument that the promulgation of his views would be for the public benefit. He has therefore yielded a reluctant consent to become the victim of an interview, for which he is entitled to the goodwill of the public and to the gratitude of the writer.

THE GREAT AUSTRALIAN "DESERT" OF THE PAST.

Some doubt has been expressed, Mr Tyson, of the capacity of the interior of Queensland to sustain a large population, or even to maintain live stock in health and to the profit of their owners. Will you kindly favour me with your opinions?

I am reluctant to appear in print, but I have very definite opinions upon the inland country of Queensland, and of Australia. I think that no country in the world is so well suited to for the propagation of animal life in its highest forms. Experience shows that whether it be developed in sheep, cattle, or horses, the vitality of animals born in the interior is "greater than anywhere else. The true elixir of life, whether in the form of air or food, enters the system by the throat of all animals. Thus, our dry, pure, rarefied air, our genial sun, and our boundless space combine to promote the development of animal life in its most vigorous and perfect forms. The rather scanty rainfall, with the almost incessant sunshine, produces sweet and nutritious grasses and herbage such as are found in no other part of the world to my knowledge. Even the physical health of the animal is conserved as its natural functions are stimulated by the great expanse of country usually within the line of vision. Body and mind are both cultivated through the eye. The noxious qualities of the vegetation, the rancidity of grasses, and the unwholesome pungency of the herbage are purged and evaporated by our fervid sun, and by the comparative absence of humidity from the atmosphere, while our spirits are lightened by the dry and exhilarating climate. This is of course the inland climate, and not the sometimes depressing atmosphere of the coast.

Is this a mere theory, and does it apply also to the human animal, Mr Tyson?

These are not mere speculations or theories. With due care the offspring of all animals, human or inferior, bred in our inland country, are of a higher type than their progenitors. The men of Central Australia will be among the picked men of the world, physically and mentally; the horses will be the fleetest and most enduring; the cattle and sheep will yield the most tender, succulent, and toothsome meat. England is little other than a manure heap whose soil is poisoned by the animal droppings of centuries, and whose vegetation is at least less wholesome for stock than the health-giving native grasses of inland Australia, whose pastures are constantly purified by a tropical Sun and occasionally renovated by bush fires.

THE AMELIORATION OF DROUGHT

But is not the capriciousness of the rainfall, with the frequent devastating droughts, a terrible drawback?

The droughts are discouraging, but we are ameliorating their effects by artesian bores and surface reservoirs. This (presenting a quart wine bottle of sparkling water) is a sample from the ninth bore just put down on my Tinnenburra run, which was previously destitute of surface water. It is beautifully soft water, and the bore yields between three and four million gallons a day.

Are the other eight bores equally as good?

Well, I don't like to say how much they throw up altogether, but it is an enormous quantity. Some of them have been running for two years without noticeable diminution of the flow. The fall of the country is slight, and there are lakes of water as much as seven or eight miles long on the run. But we have absolutely no grass within reach of the water. I have spent a great many of thousands of pounds on these bores, but must spend a great deal more in distributing the water over the run to utilise all the grass of the country.

Have you got artesian water on any other of your runs?

I have just sent out a party of eight brothers to my station, Glen Ormiston, in the far north-west beyond Boulia, with one of the best well-boring plants in Australia. It is equal to sinking thirty bores, and was manufactured in Brisbane under superintendence of Mr Woodleigh, a well-known expert, and has just been forwarded by carriers from Longreach on the Central Railway, to Glen Ormiston. The plant has 16,000ft. of tubing, it weighs 100 tons, and the carrier's contract is at the rate of £16 per ton.

These improvements are very expensive, I presume?

Yes, I have spent many thousands of pounds at Tinnenburra on bores. The Government, by the way, reserved all the natural water on the country and owing to the prolonged drought and the disappearance of grass within reach of stock the artesian water I have obtained is at present almost useless. A great deal more money must be spent in its distribution, and now, just as my ninth bore is completed, the Land Board has served a notice of revaluation of the run: which means, I suppose, that I am to be rack-rented on my own improvements. The Glen Ormiston run has given me no return whatever for five years, and the well boring will be very costly. Another northern station has never paid me a penny; of course, they are not all so; I have so many stations in all the colonies that some of them are always paying. The several governments of Australia get from £18,000 to £20,000 a year from me in rents.

Is your well-boring done by contract?

Yes, all by contract. (Looking at his note-book); I find my contract with Thomas Hannay Bros. For Glenormiston prescribed a rate per foot varying according to depth, beginning with 10s, running into pounds at great depths. I find all plant and machinery, supply rations at actual cost with 10 per cent added for carriage and beef at 1d. per lb., the contractor to fetch it from the station to his camp. In the event of the contract being faithfully carried out, however, I remit the value of the beef as a bonus. On these terms the contractor is to sink as many bores as I think fit.

You must be heavily burdened with the management of your many properties in the different colonies, Mr Tyson?

Yes, I am a busy man, but I have good overseers and managers. It has always been my custom to train my men myself – break them in as boys. I can then thoroughly trust them. I am very strict, you know. No swearing allowed on my stations, and any man who indulges in vehement language at once gets his cheque. No alcoholic liquor permitted

— I never use it myself – no card playing. To bed soon after sundown, and up at sunrise. That is my habit, and all do as I do.

But you have no amusements for yourself, men or boys?

No, life with us is a serious business. Reclaiming the country is real work.

HOMES FOR MILLIONS OF PEOPLE

What is your opinion about settlement in the interior – do you think the country is fit for cultivation?

I think that farming settlements should be created where the land is suitable all through the country. Farms should be laid off on the river and creek frontages where the land is good. The frontages should be held by the farmers and graziers of small means, and not given to the wealthy squatters as has been too often done in the past. The men of large means should be allotted the back lands, which they should use their capital in developing by sinking for artesian water and making surface reservoirs.

But there are two difficulties, Mr Tyson. When halves of all runs are resumed everybody is served alike, but if spots were to be picked out here and there, where, as you say, the land was suitable, some squatters, unless compensation were given, would suffer greatly, while others would profit by the increase of settlement. The other difficulty is how to define large and small holdings.

I admit compensation must be given, and there is the liability to abuse to be faced. But I believe that if competent valuers, not mere collar-and-tie office men who know nothing about the country, were sent out to appraise the land surrendered at its honest value the difficulty might be got over. As to the size of holdings, that is a great problem. If you ask my opinion, I would have freehold tenure, and give every man so much land as he had means to properly work or develop. The land should be put to its natural and reasonable use, not held for speculation, the circumstances of the country for the time being considered, and on that principle every man might get as much as he could use. I hold that land must be set apart for different purposes, and that the Robertson Act of 1861 was the worst Act ever passed in Australia. It made selectors for the purpose of plunder, and not for founding homes or developing the land. Fancy selectors taking up areas between a squatter's homestead and his run for the express purpose of cutting him off from his country and compelling him to pay blackmail by buying the selections at a high price! No, free selection everywhere up to 640 acres under that Act was a curse to the country, and bred class hatred and strife. It is for the benefit of the squatter to have farming settlement in his neighbourhood, and there are places all over the interior where men of such means could settle, and where fodder, grasses and plants could be advantageously grown, and fruit produced with profit.

Then you are a believer in close settlement in our inland country, Mr Tyson?

Undoubtedly, I am. I am confident that our productions could be enormously increased, and that for every bale of wool now raised we shall in the future produce 1,000 bales, and the finest wool in the world. There is room for millions of people in our inland country. But the Government must not oppress the people on the land by taxation or by rack-renting if they wish the country to prosper.

THE BORDER AND STOCK TAXES

Is the taxation in Queensland really oppressive, Mr Tyson?

Certainly, it is. Take the case of South-western Queensland now. The country is stricken and paralysed by drought, by the low price of cattle and wool, and by carriage difficulties. I admire Sir Thomas McIlwraith in many ways, but I think his meat export tax and his border tax perfectly monstrous. I am sure the far inland squatters cannot pay them, and if they are enforced the unfortunate men will be crushed. It will be the last straw. Why, you can get 10s. a head net for Queensland cattle driven to the Murray — in fact, even that price can't be got now. The meat bonus will be of no use to any but the coast squatters.

But I heard at Charleville the other day that the New South Wales Railway Commissioners were paying the border tax on wool, and that of the stations that formally used Charleville two (sic) were this year sending to Bourke. Is that true?

Well (smiling), I know all about that, but prefer to say nothing. The people can't pay the tax, that is certain.

COLD STORAGE INLAND

Why, Mr Tyson, will the meat bonus be of no use to the inland squatters — have you no faith in frozen meat — surely it is better even at present rates than 10s. a head at Albury?

Oh, yes, I believe in frozen meat; I am sure that if we could kill on our western pastures, and skip to Europe, we should send them the best meat in the world, but I will have nothing to do with meat works — it is not my business.

But, Mr Tyson, don't you see how the export of frozen meat would raise the value of cattle and sheep, and how enormously you would benefit as a great stockowner?

Oh, yes, if the works were taken up by the right men.

But surely the right men can be found? And surely, with the factory bonus provided by the meat tax, the necessary capital would be forthcoming if the stockowners only took the matter up and found half the money?

I will have nothing to do with it. My business is grazing and I already have as much as I can look after.

But what is the use of raising multitudes of stock if there is no market for them?

Oh, I can wait for my market, or — well, shoot them.

But in a national and patriotic aspect, is not the establishment of inland meat factories, with the feeding of the ill-supplied people of the mother country, an inducement to you to assist in this enterprise? As a patriotic Australian, do you not think it would be a grand thing to assist the frozen meat enterprise with a small contribution from your capital?

I am anxious to do good to the country, and have unbounded faith in its future, but at present I will not touch the frozen meat industry. We have not the right men yet.

Well, Mr Tyson, I am grievously disappointed at your decision. Cold storage can only be economically provided on a large scale, and I am deeply impressed with the conviction that a huge cold store inland on the Southern and Western railway line would be an incalculable national benefit. Such a store situated at Gowrie Creek, close to the coal mine, would serve all the country near the railway lines for hundreds of miles to the west and south. Slaughtering places could be served anywhere within twelve hours' journey by goods train, and the primest cattle and sheep could be killed close to their pastures and the meat at once stored. The, whenever there was a vessel ready in port, it could be quickly run down by rail. Do you agree with me in this?

Yes, I think it would be a good thing, but I won't touch it.

ROADS AND RAILWAYS

Well, Mr Tyson, you complain of the border tax, but don't you think Mr Eddy and his brother commissioners acted very improperly with their differential rates?

I think Mr Eddy is anxious for justice and fair play, and is personally desirous of assisting rather than injuring Queensland and the same of the Victorian Commissioners. It is a shame that the politicians should insist on the suicidal policy of competing railways. All the railway systems of the colonies should be worked as one, for the benefit of all. I am sure the commissioners would do this if they were permitted. They have just done one good thing at my suggestion. The loss in driving Queensland cattle to the Victorian border is about 300lb. On each beast – that is, we only get 500lb. On the Murray from a beast which on leaving on leaving our own border weighed 800lb. I represented this to the Railway Commissioners, and that their charge from Bourke to Albury for trucking, particularly in the case of store cattle, was prohibitory (sic). After conference between the Sydney and Melbourne Commissioners the rate was reduced by 20 per cent, and stands now at £1 per head for lots of not less than 850. That will be an immense boon; and I hope to get a further concession in the form of permission to send cattle to either Narrabri or Bourke. The difficulties of travelling stock are increasing every year, and the recent sale of land near Bourke in lots of 10,000 acres, with frontages to ten-chain roads, has done much to make the terminal accessible to travelling stock. These ten-chain roads have now become mere dust-heaps and greatly impede the stock traffic on the railway. If, therefore, access was given by the Commissioners to Narrabri that place would be easily reached from Goondiwindi via. Moree along a generally well-grassed route. From Narrabri they could get through by rail via Blayney and Young, instead of as now from Barringun and Ford's Bridge to Bourke and thence to Albury.

In conclusion Mr, Tyson said he had strong views on the question of roads and railways. He thinks the Government are (sic) in too much hurry in these matters, that the numerous unprofitable branch lines are a striking object lesson, and that even the Charleville line has been run in the wrong direction, to the grievous wrong of South-western Queensland. The authorities ought to be satisfied, before making a railway, that there would be traffic upon it to make it pay. As a native Australian he deplored the rivalry, ruinous competition and mischievous quarrels that were being carried on between the colonies. He hoped to live to see the time when they would all be under one government, and the railways run and taxes imposed in the interests of the whole country, and when local and inter-colonial antagonisms and jealousies would altogether cease.

With the utterance of this patriotic aspiration the interview closed.

Brisbane Courier, Friday, 8 Dec 1893, P 4.

THE TYSON INTERVIEW

THE interview with the Hon. James Tyson, which we publish this morning, contains several remarkable disclosures of opinion on practical subjects. Everybody in Australia has heard of Mr. Tyson; very few have any personal knowledge of him. He is a picturesque and yet indefinite personality. He was born in New South Wales, has never travelled in the oversea sense of travel[5], is seventy years of age[6], and has just been persuaded to accept a seat in the Legislative Council of Queensland; but he has lived his three score years, for the most part, in comparative solitude, unseeking and unsought by society[7], devoting a marvellous energy to the pursuit of wealth, mainly through one industry, and glad to keep outside the range of impertinent curiosity while he extended his far-seeing operations from colony to colony, and, on continental breadth of foundation, built up his massive fortune. A strong, shy Central Australian bushman is to-day the richest man in Australasia.

We could easily wish he were a different man. A millionaire without amusements, without vices, and without humour strikes us as a strange character. We could wish that Nature had placed him constitutionally a little nearer to his ordinary neighbour, except, of course, in the matter of his neighbour's vices. But it is idle to quarrel with Nature's ways of fashioning strong men. We should be glad

---

[5] There is evidence that James went to England and New Zealand as described later.

[6] James was born on April 8, 1819 making him 74 at the date of the interview: Zita Denholm, *T.Y.S.O.N.* Triple D Books Wagga Wagga 2000, pp.21, 30.

[7] James was invited several times to Government House but he declined.

enough to have them in any shape. All sorts of criticisms have been and will be poured upon Mr. Tyson's fortune; the average man is poor and objects to individual accumulations which take astonishing figures to express them; but the power that accumulates, not by finding a vast deposit of precious metal, but by steady, bold, and intelligent enterprise, mostly self-initiated, compels admiration. Mr. Tyson modestly claims that he has spent his life in reclaiming country, the dry interior lands of Australia, and that this is real work. "Life with us is a serious business," he says. "With us!" Who are the others? "I am a busy man, but I have good overseers and managers. It has always been my custom to train my men myself – break them in as boys, I can then thoroughly trust them. I am very strict, you know. No swearing allowed on my stations, and any man who indulges in vehement language at once gets his cheque. No alcoholic liquor permitted – I never use it myself. No card-playing. To bed soon after sundown, and up at sunrise. That is my habit, and all do as I do." It would be hard, perhaps impossible, to get the equivalent of this from the lips of any other millionaire in the world. It is one of the oddest bits of puritanism on record. And the sincerity of the utterance is not diminished, though the generous reader refuses to press it absolutely. The overseers and managers, disciplined from boyhood, may have adopted and adored the Tysonian habits, but we fear that the potent whisky, the vocabulary of passion, and the seductive card have not vanished from the Tyson stations.

Mr. Tyson is an enthusiast in regard to the capabilities and destiny of Central Australia. He has spent most of his life there, he has watched with his own eyes the whole course of its occupation, he has been and is still a pioneer among pioneers, he has felt every irregularity of the climate, he knows the country in every detail as few other bushmen know it, he has made most of his money there, and his heart and faith and hope are there; and now he comes among the coast-dwellers and says that Nature, so often misunderstood and blasphemed, will yet justify, nay, is justifying herself in Central Australia. Eulogy, perhaps exaggerated eulogy, is the dominant note of the Tyson view of the inland plains. The scanty rainfall, the incessant sunshine, the dry pure air, combine to promote the development of animal life in its most vigorous and perfect forms. "With due care," he says, " the offspring of all animals, human or inferior, bred in our inland country, are of a higher type than their progenitors." Central Australians, he predicts, will be among the picked men of the world. We believe he is not far out in that opinion. The old theory of the second-rate scientific institutes, that the native Australian was certain to become a mere human slab, has been falsified already. Men bred where the greater part of Mr. Tyson's life has been passed are not fleshy creatures; but, if lean, they are lithe, upright, of wholesome blood, and splendidly strong. As to stock, " England," says Mr. Tyson, " is little other than a manure heap whose soil is poisoned by the animal droppings of centuries, and whose vegetation is at least less wholesome for stock than the health-giving native grasses of inland Australia, whose pastures are constantly purified by a tropical sun, and occasionally renovated by bush fires." Mr. Tyson's view of England will be much disputed, but he puts the truth about inland Australia with great force. And he holds a bright faith in its future. He sees no impossibility in the hope that millions of people will yet be settled on the hot but healthy and fertile plains. It is gratifying to find, in this connection, that our wealthy bushman is not an enemy of agricultural or other dense settlement. The New South Wales Robertson Act of 1861 he denounces as the parent of plunder rather than of settlement, a view from which few well-informed persons will dissent, but he believes in providing facilities for small men to take up land, and goes so far as to advocate that river and creek frontages should be held by farmers and small graziers, leaving the back lands, to the capitalists who can sink for artesian water and make surface reservoirs. Let no man say that squatters are prepared to give up the frontages (on terms of compensation) because they have suffered heavy losses from floods and prefer the safer tanks and wells.

It is his personal relation to the pastoral industry that we see Mr. Tyson in the most interesting light. He brings to bear upon it an immense fund of self-gained experience, his observation of it extends over several colonies, and he is able to improve his stations in every way that seems to him a profitable investment of money. He attaches an enormous value to the recent discoveries of artesian water. On Tinnenburra, a station in Southwest Queensland, he has nine bores, and the ninth yields between three and four million gallons a day. The aggregate yield of the nine is not stated, but we are told that "there are lakes as much as seven or eight miles long on the run which was previously destitute of surface water." Lakes or streams of this size formed by artesian water, or of half or quarter this size, ought not to exist on any run in Queensland, unless it is impossible to shut down the bores. Though it be granted that after two years there is no diminution of the flow, it is certainly prudent to economise this as yet unexplored and unmeasured supply of what in western country is literally a precious fluid. However, the water is at Mr. Tyson's disposal, and as there is now absolutely no grass near the bores, he proposes to spend some thousands of pounds in so distributing the water as to enable him to utilise all the grass on the run. This should prove a very instructive object-lesson, and in providing it Mr. Tyson may confer a real benefit on his fellow pastoralists. So strong is Mr. Tyson's faith in artesian water as a contributor to successful occupation of arid country, that he has equipped an expedition with a plant capable of putting down thirty bores on Glen Ormiston station, situated in the far North-west. The carriers' bill alone will be £1,800. There is a

Tysonian magnitude about this; and at Glen Ormiston, again, an instructive object-lesson may come of the enterprise. In the development and utilisation of the artesian water supply the millionaire squatter has it in his power to so employ his resources of experience and capital as to better the prospects of a hard-pressed industry, and, by doing so, to confer some solid benefit upon Queensland.

Mr. Tyson does not believe in the border tax. He says it is the last straw to crush the drought-afflicted South-western pastoralists, who are also struggling with the low prices of wool. But somebody, Mr. Eddy possibly, appears to have taken the straw off, and the South-western pastoralists are saved for this season. The antipathy to the border-tax is intelligible, for Tinnenburra is in the South-west, but what is not intelligible is Mr. Tyson's stubborn determination not to touch the frozen moat enterprise till the right men are found. The Tysonian definition of the right man may not be the definition which would satisfy other pastoralists. Mr. Tyson might want men who in addition to a thorough knowledge up-to-date of the meat-export business have unveracious tongues and uncorrupted palates, are ignorant of the meaning of the strange signs on gambling cards, and go to bed at sundown and get up at sunrise. It is a pity that all naturally clever men in Queensland have not had the advantage of discipline in boyhood on Mr. Tyson's stations, but in this imperfect world, or at least in this not quite perfect Queensland, we must compromise if we would make practical progress. The future of the pastoral industry apparently depends in no small degree on the profitable production of beef for foreign consumers. There must be some initiatory losses, and despite the last straw of the border tax we humbly submit that Mr. Tyson is not too much crushed to join the other pastoralists in establishing this enterprise on permanently remunerative foundations. We have only a very remote, an imaginative, idea of what it is to live up to the possession of millions, but we believe that there really is such an obligation, and we should like to see Mr. Tyson discharging it with artistic and generous fulness in regard to the gallant endeavour the Queensland pastoralists are making to find a market for their beef in the other hemisphere.

In conclusion, we are glad to range ourselves alongside Mr. Tyson and all who with him heartily desire to see all intercolonial taxes* and jealousies and separateness of feeling and aim perish for ever before the incoming of an august and invincible and everlasting Commonwealth of Australia.

I know James would, have read this, because after James' death there was a trial to determine James' domicile for stamp duty purposes (as detailed later in this publication). The editor of The Brisbane Courier at the time of the interview, Charles Buzacott, testified as a witness under oath. A summary of that testament was recorded in The Brisbane Courier as follows:

***Brisbane Courier,* Friday, 16 Nov 1900, P 7.**

Tyson gave him (Charles Buzacott) an interview which was published in the " Courier" of the 8th December in that year. Five or six weeks afterwards Tyson saw him, shook him cordially by the hand, and said, " It's all right, mister; I believe it's done good; I believe it's done good." He had not met Tyson in the meanwhile. Tyson had granted the interview reluctantly, and appeared up to that time to avoid witness (Charles Buzacott}.

…

There were so many good attributes describing James in the preamble that I thought, "this can't be a relative of mine! I don't personally know anyone who would warrant such a eulogy. He could be re-classed as a Saint! I cannot believe any person warrants the accolades bestowed on James.

The replies in the interview were very eloquent for a man who had little education! And he was quite chatty. He seemed as though he loved talking. If I just read the interview, I would have felt he was enjoying himself and he had no qualms in stating his own opinions.

The interviewer, in the preamble, painted him as, "...somewhat embarrassed by the shyness of a bushman unaccustomed to contact with civilisation." And later, "Naturally of a retiring disposition, ..."

Towards the end of the interview, the interviewer showed a hidden agenda being to induce James to get behind the push to export frozen meat to England; and was a little piqued at James' refusal! He did not hide his abhorrence at James' refusal citing, "unless the right men are engaged to run it." The Editorial pressed this point further as an attempt to shame him.

To put the record straight, James was speaking from experience. I found this report of a meeting in the Brisbane Telegraph. He was a member of a steering committee looking at the exportation of meat to England 15 to 20 years prior. A meeting was held on December 6, 1878 and below is an extract from the report of the meeting.

Telegraph, **Saturday, 7 Dec 1878, P 2.**

THE EXPORTATION OF MEAT.

The Chairman, in opening the meeting, said there are two gentlemen—Messrs Tyson and Beardmore—present who will lay before the meeting the particulars of the Chicago process being successfully carried out in Melbourne. He therefore would call upon Mr Alfred Brown, who was best acquainted with the frozen meat question.

... (Mr Brown then described his version of the freezing process.)

These terms did not prove satisfactory to the meeting, and were regarded as impracticable. **Mr. Tyson then described the manner in which the Chicago system was practised in**

Melbourne and some tinned corned mutton was tasted by the company and pronounced excellent. The universal opinion was that the absence of stringiness formed a marked improvement upon anything previously produced.

Mr. Beardmore said the meat just tasted was, he thought, better than the Chicago sample, which seemed to be ground up, and was rendered less succulent by its dryness. He did not think any machine method of compression was used in Chicago. A workman lately employed there had told him that the meat was put in by hand, and the compression resulted from the collapse of the tin cooling. He thought the Chicago tins were better, being tapered.

Mr. Tyson said a gentleman had written from England to say that the Kent had just arrived in London with a shipment of Melbourne meat, and it was there worth 61/2d. to 71/2d. per lb., beef and mutton respectively. He had been assured that these prices would pay well. Mr. Davenport, M.L.A., considered that they should leave the dry air freezing system alone, until a trial shipment had demonstrated its success. He thought the meat-preserving work already in the colony might be utilised, and believed that the Oakey Creek establishment might be secured at a moderate price if a company were formed to buy and carry it on.

The Chairman believed those works could be bought for about £7,000.

Messrs. Morehead and Wienholt were in favour of securing an immediate market, maintaining that it would be better to boil than to do nothing.

Mr. Brown dwelt upon the advantages of compression; it being said that 14 lbs. of meat could be condensed into a 7 lb. tin. Thus, tins and freight would be saved.

On the motion of Mr. Davenport, it was then resolved that a company be formed, with a capital of £25,000, in £1 shares, to carry on meat-preserving in this colony on the Chicago Melbourne process; and the motion having been agreed to, shares representing £3,500 were at once subscribed in the room.

> Mr. Davenport then moved that Messrs. Bell, Barker, Morehead, Walsh, Brown, **Tyson**, and the mover, be appointed a sub-committee to make enquiries, draft prospectus, and report to a future meeting. The motion having been adopted; it was arranged that the sub-committee meet at four o'clock in the afternoon. Mr. Brown said that what was needed was the use of the patent.

This quotation tells me two things about the man. One was the way James exerted his influence over his peers and they responded by accepting all he said. Similar to the interview where he made sure his viewpoint was conveyed in a very definite manner. There were many years difference between the two events but, James still made sure his thoughts were heard in a similar manner.

In terms of my quest to learn more about my uncle the interview did little to help me, apart from his attitude to the human and animal product of Australia in general and Queensland in particular; in contrast to the preamble which told me more about the man. But how accurate was it when compared to the Exportation of Meat article?

CHAPTER 3

Articles from the interview to his death, 4th December 1898

Articles in this chapter are those published before James died but after the interview; therefore, he was able to read the articles and refute anything not right. It is interesting to compare the content of the writings with those published before the interview.

The following article is reproduced exactly as it was written. That is why it has no paragraphs and is very wordy as was in the original; nevertheless, I believe it warrants inclusion for the new matters it brings about James. Morehead was James' agent so Dr Grant would have known him well.

The Week, Brisbane, Friday, 15 December 1893, P 7.
Hon. JAMES TYSON, M.L.C.
A Character Sketch
Dr A. C. GRANT. of Messrs. B. D. Morehead

The inevitable interviewer has at long and last run Tyson to earth. " Who would have thought it," some will say. Who would have believed that Tyson, the quiet, retiring, self-contained, and somewhat cynical lord of cattle upon a thousand hills, would ever have delivered himself into the hands of the Philistine interviewer? Others, with a deeper insight into human nature and with special opportunities for observing, will aver that a certain amount of notoriety has always proved acceptable to him, and increasingly so with advancing age. Be this at (sic) it may, there are very few who do not wish Mr. Tyson well, or who do not admire and respect the stern, indomitable power, of will which has enabled him all his life, on his own particular lines, to soar above all other Australians. To all colonists the straight, unbending, one-ideaed (sic) man will for long be one of the most picturesque figures of the century, not only as the possessor of what may almost be termed boundless wealth, but for personal characteristics as rare in some particulars as they are admirable. To say that Tyson has no faults would be to contend that human nature is perfect. No doubt he has faults in common with other men, but he differs from them in this respect, that while the sayings and doings of most other men may be, and often are, passed over as comparatively unimportant, the doings and even the casual remarks of so great a power in the money world as he is are invested with a significance possibly never contemplated by the speaker, whose peculiar mode of life and earnestness of purpose leads him often to employ expressions and illustrations as strange and novel as they are unintelligible to the ordinary man. The well-known simplicity of one who, having it in his power to command nearly all of the objects in life for which men toil and scheme, yet prefers to continue to pursue with unremitting determination and vigilance the task which has animated him from early life has been found to be an excellent basis for stories of his eccentricities none the less amusing because they are "bcu trevalo." (sic) Those who may regard him as an Australian edition of Jay Gould are however, vastly mistaken in his character. Never were two men more dissimilar. Above all Tyson is a patriot. No one who knows him can ignore his passionate devotion to the land of his birth or his love and esteem for native-born Australians, as compared with other races. As a philanthropist he has never posed, still, great schemes of this nature, though crude and undigested, have undoubtedly flitted through his brain. His known charities may not have been so extensive as they ought to have been in the opinion of those who possibly think they could manage his affairs better than himself, but it can never be said that, his purse remained shut or his ear closed to any bona-fide appeal for assistance. His private charities will never be known. Ostentation can never be laid to his charge, but this much is certain, that when appealed to by those in whom he had faith, he has never remained callous or indifferent to the requirements or sufferings of humanity. Far from it. It must also never be lost sight of that to a man of such concentrated powers of accumulation it must be a difficult matter to part with anything, and possibly were it not for the strong moral principles which govern him he would develop into a sordid miser. To this stage Tyson has never descended. He accumulates because his nature drives him to accumulate. He saves because his wants are very few. He continually acts up to the maxim that " 'tis not what we make that makes us rich, 'tis what we save." Look at him striding down the street with his thumb in the armhole of his waistcoat and his head in the air, a good foot above most other people. His clothes are well cut and of good dark cloth, but they have seen good service, and will probably owe their wearer nothing by the time they are discarded. A soft felt hat, a Crimean shirt, and paper collar complete his costume, while a thin strip of leather or greenhide does duty as a watch guard. The paper collar and greenhide watch guard are merely eccentricities allowable to a millionaire. They do not betoken miserly saving or sordid avarice in the smallest degree, but are rather protests against the extravagances of the age. His whole life has been devoted to protesting against luxury. In his appetites he has always been a marvel of abstinence. He does not smoke, and drinks water only, or at most ginger beer. He can make a meal of a perfectly satisfactory nature of a pie-melon or a crust of bread or a few apples. He will camp down at the creek instead of using the adjacent public-house. He will live contented for as long as is necessary on damper and salt junk. Formerly this love of simple diet was far more pronounced than it is now when he has reached the allotted span of man's existence. In all probability he occasionally feels now that a well-cooked square meal is very acceptable, but he is as rigorous as ever in his tirades against extravagance and luxury. The young man who parts his hair in the centre, or who uses scent or any other specimens of what he terms the "la-de-da " class, are especially obnoxious to-him. On the other hand, hard-working, capable, and trustworthy men find in him a good master, more especially native-born Australians. As a thorough bush man himself, he is capable of forming a correct estimate of work done, and he is equal capable of showing his appreciation of the same. Not long ago, on a most unexpected manner, he left a small wooden box to be delivered to one of his drovers who had completed a good trip south with cattle. Upon, being opened the case contained one of the best gold watches obtainable with chain complete, and better still, on the inside case an inscription testifying to the merits of the drover. It must not, however, for one moment, be conceived that he is easily assailable for money. Probably no more difficult task could be found than to endeavour to get him to invest money in any of the thousand and one methods of losing it adopted by other men. If he understands anything, he understands taking care of what he has gained. He well remembers what it cost him to make a start and to lay the foundation of his fortune, and when in a communicative mood will readily narrate his hardships as a boy tailing cattle, and his disappointment as a young man when he found that his twelve month's work had had been paid for by a valueless cheque. It will, perhaps, interest and amuse some to know that the foundation of his great herds was a brindled cow, which, after much solicitation and debating on through an entire night, he took in exchange for an accordion, which he had purchased on completing a droving trip, and with which he was solacing himself on his return journey. The rapid increase from this veritable " milky mother of the herd" turned his attention to the acquisition of stock, and to this pursuit he still continues faithful. To follow the Australian millionaire through the vicissitudes of his hard-working life is not necessary. He had his early ups and downs, but a steadfast, unalterable, inflexible purpose enabled him to overcome every opposition. What boots it (sic) to tell how the country which he lost by a piece of injustice in his early start now forms part of one of his stations. What need of narrating how he purchased large lots of sheep and cattle during the early days of the gold field, and after travelling them through long stretches of unoccupied country and undergoing privations from hunger, so remarkable and severe as to earn from some, who supposed that he could only have subsisted on herbs, the name of Nebuchadnezzar, sold them at almost fabulous profit. These are merely incidents in a career characterised by courage, strength of character, thrift, and exactitude in business which may well serve as a model to his fellow-colonists. The personal appearance of this very remarkable man is no less striking than his history. Over 6 feet 4 inches in height and as straight as one of our native gumtrees, splendidly built, with a sinewy nervous, spare, but cast-iron frame, he was indubitably framed by nature to undergo hardship without discomfort. In particular the head is exceptionally handsome, and would be singled out for admiration in any assembly in the world. The cranium, full, round, and clean-cut; the eye blue, clear, steadfast, piercing, and particularly noticeable when lit up by the fire of enthusiasm; the nose thin, rather bony, but exceedingly well formed; the lips firm and well cut, shaded by a moustache which, with the handsome short, thick, and somewhat curly beard covering the lower portion of his face, as well as his hair, is of a silvery white— in short, a finer or more aristocratic head it would be hard to find in any country of Europe, apart from Australia. Further, the dignified appearance and carriage of the man have something essentially peculiar to himself about them, and the resolute, decided, commanding, yet deliberate and very wary look which sets so well on his bronzed-marked features, gives him the air of a general of division employed on active service. On the other hand, his pure life, his quaint expressions unblemished by the faintest approach to looseness or profanity, his intense lore of nature, his wonderful individuality, combining the simplicity of the dove with the wisdom of the serpent, and the unmistakable suggestion which he carries with him of the great western plains, irresistibly remind one at times of Fennimore Cooper's famous hunter "The Leatherstocking." He is a member of several good clubs and frequents them, but under no circumstances can he be described as a clubable (sic) man as generally understood. His absorbing devotion to business, involving the continual performance of long and arduous, journeys for the purpose of personal inspection and supervision, allows no time for relaxation or social amusement. Life with him is indeed a serious business. It is the accomplishment of a mission. In fact, the originality of his character is only equalled by his natural grasp of finance and mastery of detail. Although nearly 70[8] years of age, he is so far like Moses that his eye is not dim nor is his natural force abated. At the present moment he is engaged with the whole force of his nature in converting immense areas of arid pasture into a land intersected by running streams of water, which one day will be tenanted by thousands of the people of the race he loves, and, if spared, much more may be expected from him in the same

---

[8] James was born in 1819 so he was 74 at the time of writing

direction. Of course, it may be urged that he is benefiting himself by his action, which is true. It is none the less true that he is doing a great work for the colony, and at any rate it can never be urged against him that he spent as an absentee the money he accumulated in the country. In concluding this sketch, it may be just as well to intimate to the fairer portion of the community that, its subject, although a bachelor, does not contemplate matrimony, for the reason which has kept him single all his life. He has not time to marry. Marriage would divert him from the aim in life which he has set before himself. Still, there is no greater admirer of virtuous, healthy womanhood in the world, nor anyone who feels more the necessity for inculcating the obligation of work and thrift among the young. But it may be naturally asked, "What does he propose doing with his accumulations? "Who knows, indeed? Perhaps some great philanthropical scheme is taking shape in his brain even now in connection with his native land. He has already given evidence that when required he can do a patriotic action, as witness his subscription to the Soudan affair, and his purchase of Queensland Treasury notes. Perhaps he does not well know himself, but one thing is certain, no one on this earth is more impressed with a sense of responsibility in regard to wealth. Meantime he is one of the cheeriest of mortals and extremely fond of a joke. Who of his acquaintances does not enjoy the mysterious manner he adopts occasionally when endeavouring to impress his hearer, and more especially when, in reply to the question, " Hello! How are you getting on?" he lowers his voice and confidentially, whispers, "Just keeping my head above water, mister."

A quick note from the "Father of Federation" is a welcome change after the above!

Toowoomba Chronicle, Saturday, 30 Jun 1894, P 3.

NO TITLE

Sir Henry Parkes has a very high opinion of the Hon. James Tyson, M.L.C. In the course of an interview with the Sydney representative of the Courier on Sunday last, the politicians of Queensland were discussed, and Sir Henry interpolated his opinion of Mr. James Tyson: "I see he has gone into politics. It was a surprise to me when I learned that he had accepted a seat in the Upper House in Queensland. A seat in the Council here has frequently been offered to him and always denied. Mr. Tyson is one of my oldest friends. When I kept a small warehouse in York Street many years ago, he often came to have a chat with me, and a mutton chop. He was a great talker, and would give me his opinion on geology, and more especially in its relation to the constitutional elements of the soils to be found in various parts of the Colony. Mr. Tyson is a great reader, and a man who discriminates. He used to come to me and get a list of useful books, which he would then immediately purchase. He knew more of Darwin's works than possibly any man in Australia. No man, I think, is more misunderstood than Mr. Tyson. He is an entirely unselfish man, simple at heart, and pure in life. Why he went into politics is a mystery which I cannot solve."

The Telegraph, Thursday, 29 Nov 1894, P 2.

Hon. J. Tyson
Interviewed at Charleville.

The Hon. James Tyson, M.L.C., arrived by Wednesday evening's train, and started the following morning for Cunnamulla, on a tour of inspection of his southern Queensland properties, thence he works his way to Hay. Our (Charleville Times) reporter had a short interview with the gentleman. Asked his opinion of the rabbit question he replied:

Mr., the people of Queensland know nothing about rabbits. They have been a curse elsewhere, and will bring desolation in Queensland. Nothing but fencing, and plenty of that, will combat them. And let me tell you, there is another curse in store for them— the prickly pear. There ought to be legislation in time, if not to eradicate it, check its terrible spread.

What do you think of the increased rental the squatters will soon have to pay?

It's wrong. The system is not the right one. Instead of asking the lessees to pay a tax on their improvements on leasehold, they ought to improve their holdings. I will illustrate what I mean. I have country, leasehold with frontages to never failing water. In a dry season there is no grass near the water and no water near the grass. The Government should send their man out and order me to put down a bore, or dam, or tank, as the case may be. The landlord, the Government, would then be getting his land improved, and at the same time getting an in direct revenue from the increased stock it would carry.

Why don't you take a rest, Mr. Tyson?

Mr., I'm going to work to the end, it is my life.

Do you never weary of the constant travel?

Well, my last trip to the Northern Territory was the worst

I have ever experienced — that is, it has taken more out of me. You see there was no defined track for the greater part of the way; we followed travelling cattle tracks for days; the cattle had travelled in the wet, and this was very nice in a buggy. We had to carry preserved meats and fish. The roads being rough, the preserves got churned up, and we had ultimately to live on preserved soup. It was awful. I have not got over it yet.

Now, Mr. Tyson, I seldom see your name in the debates in the Council. How's that?

Right, Mr., I am getting deaf, but never properly know it till I entered the House. I can't hear a word when the House is sitting. In committee I am of use, and give the members the benefit of my experience, which they find very useful.

How is it that your name never appears in the railway books as a consignor of wool?

The reason is simple, I can despatch my wool with greater certainty from my South-Western Queensland properties via Bourke than via Charleville. There are two creeks, the Tuen, and the little Tuen, between here and Tinnenburra, they are the trouble. If these creeks were bridged my wool would come to Charleville.

How would federation suit you?

Now you have struck my weak spot. I am an out and out federationist (sic). I consider the colonies should trade as one. We are the same race, have the same wants, share the same privations, enjoy the same clime, and yet we are vying with each other in carrying our staple to the seaboard.

It was suggested to me that I should not include the next quotation. I was advised it is too long and the authenticity may be suspect. I thought long and hard about it and decided it was an important document because I counted twenty-three items about James, with most not advised elsewhere. Additionally, it was published by a major Brisbane paper when James was still alive so he could have read it. See if you cand find more than twenty-three items to help us understand what James was like or what he did. The most poignant fact was his droving cattle. On foot! Then quietly moving through the groups to count them. So different to the normal drovers with their stock-whips and barking dogs and horses to keep the mobs together. James obviously had a special affinity with horses and cattle.

The Queenslander, **Saturday, 20 January 1894, P 118.**

Sketcher
Three Days with Tyson

About five years ago I travelled North in the Arawatta. It was growing dusk as she left the wharf, all good-byes had been said, and as the supper gong was sounding most of us picked up our travelling traps and dived for our cabins. In my cabin there were two bunks; on the lower of these lay a tall man, with coat and waistcoat off, and a red handkerchief bound around his head by way of a nightcap. I bade him good evening. He nodded, and then in silence watched me as I unstrapped my bag and got out some things. Seeing me unfold my pyjamas and throw them on the top bunk he asked: "Are you going to turn in now?"

"No," I replied, "not quite; but you look as if you had gone to camp for the night."

Ah!" he said. "You're a bushman, I see. I thought so at first, but the travelling bag and the pyjamas puzzled me a bit." And he looked proudly down at an old worn saddle valise and a rug rolled up in a tanned kangaroo hide.

"Do you think, then," I observed, "that no bushman ought to own a bag or wear pyjamas?"

"That depends. What are you?"

"A drover."

"Cattle?"

"Yes."

"What! with a Gladstone bag! And pyjamas to wear at night!" And he laughed at the idea.

"Why, you'll require a dray and horses to cart that along, let alone your blankets."

"Well, of course, I'll have a dray and horses to take our swags and rations in."

"And the pyjamas," he added with a laugh.

"No," I said, rather annoyed, "if you come along, you can have them, once we start, as you appear to need them."

"Oh, then, you don't put them on when you turn in, and wait to change when you're called for watch?"

"Well, not exactly; when I'm called, I pull my boots on with the spurs already fixed, and I consider myself dressed. When I turn in, if it's fine weather, I kick them off, and I'm undressed."

"And if it's wet?"

"Then we don't even take our spurs off; we might catch cold."

"Ah, that's a bit better; but a dray and horses. Why, when I was droving, we never thought of such a thing—nothing but a packhorse."

"Oh, you're one of old Tyson's sort—a one-horse, quart-pot[9] drover; but I'd like to see you or anyone take a thousand bullocks a thousand miles to market with a packhorse."

"I could take a mob on foot," he said, "and that's the way cattle ought to be taken, instead of knocking them about with horses."

"Rubbish," I replied. "I'd like to see you try it with a mob of Tyson's Tinnenburra bullocks."

"Why Tinnenburra bullocks?" he asked, suddenly sitting up in his bunk.

"Because," I replied, "they are the worst cattle on the road; brutes to rush."

"I'll lay you," extending his hand, "three thousand pounds to one that I can take a mob of Tinnenburra bullocks by myself on foot."

"Bosh!" I said; "you're no more than any other man; you might start with them, and see them make a start, but that's the last you'd ever see of them till you got horsemen to muster

---

[9] Billy or small tin bucket holding a quart used for cooking and drinking.

them again." With this I left the cabin and entered the saloon, where most of the passengers were sitting down to tea.

After supper I went on deck, but finding it was raining, returned for my oilskin. My cabin companion lay as I had left him, still wide awake. The moment I entered he asked me how far I was going. To Normanton? (sic) I answered that I was going to bring bullocks in at least a hundred miles from there.

"From what station?"

"G____ station."

"How many?"

"About 1000."

"How many men will you have with them?"

"Three or four with the cattle, a horse tailer to look after the horses, a cook, and myself."

"What wages do the men get?"

"Thirty-five shillings a week."

"And how much do you get?"

"A fiver."

Here he appeared to do a little mental calculation, and then said suddenly, "They can't do it, and make it pay."

"I don't know," I said, "they did it last trip, and it paid very well. The bullocks brought £4 12s. 6d. in Wodonga yards."

"I tell you, man, they can't afford it; they are all poor men out there."

"How do you know?" I said; "they might be very well off."

"Not they," he cried; "if they were, they would never have gone there."

"Who owns the station?"

I told him, and added, "They are far from being poor."

"What's the name again?" he asked. I repeated it. "And yours?"

"Mine, D____y, better known as 'Poor D____y.'"

Now, I thought, old man, you've had the handle long enough and done quite enough pumping. Allow me to give you a spell for a bit. So, I began by asking, "Where are you going?"

"Rockhampton."

"What for?"

"To have a look at a station."

"Going to buy?"

"No, I own it."

"Oh! What's your name?".

"Smith."

"Ah, I have met people of that name before. Know some

intimately, in fact." "Where did you come from?" I continued.

"From a station of mine on the Barwon, above Walgett."

"Oh, then, you own two stations?"

"I own several."

"But there's no one named Smith owns a station above Walgett; I know all that country. What's the name of the station?"

"Oh," he replied, "you are a bit too inquisitive."

"Not a bit more so than you. You asked me where I was going; what I was doing; what my name was; and what wages I was getting. Now, good night; I'm going on deck for an hour." When I came below to turn in my inquisitive and queer acquaintance was asleep.

Next morning, he was lying still, but wide awake, with hands clasped under his head, just as I had first seen him. He did not speak, but as I picked up a bath towel, I noticed he was scrutinising me—or it might have been my pyjamas—very closely. Having enjoyed my bath I returned to dress, and without a nod or good morning he said,

"Do you have a bogey[10] every morning?"

"Yes, when I can. Don't you?"

"No! It's the greatest mistake in the world."

"Well, I am sure," I said, "doctors don't think so. Why do you?"

"Because, see, it's made you shiver, and to regain the natural heat of your body will require the expenditure of a certain amount of vital power, which means so much strength wasted. Look at the blacks; they never wash themselves in cold weather, and see how healthy they are, and what soft skins they have."

I thought it was a strange view, and told him so, adding that a black was a bad example to follow.

"Not in some things," he said; "not in that, at any rate. Giving your system a shock by bathing on a cold morning means an expenditure of vital power afterwards which is so much gone to waste."

---

[10] Wash or bath

"Ah, well," I replied, "it gives one an appetite anyhow, and, as breakfast is ready, I'll go and try to recover my lost vitality."

About two hours afterwards as I was leaning over the taffrail someone touched me on the shoulder. A tall, wiry, handsome grey bearded man, with a kindly smile and a look of laughter in his keen bright eyes, was standing by my side. At first, I scarcely recognised him, but the moment he spoke I knew his voice. 'Twas my cabin mate.

"Well," he said, "what are you doing?"

"Nothing," I replied; "enjoying the sweet idleness of the hour. I delivered a mob of bullocks at Wodonga on Thursday last; this is only Wednesday following. So, you see I have lost no time."

"No, indeed you haven't; but you're hardy and used to travel, so it won't hurt you. What did you say your name was?"

"D____y," I answered. "And yours is Smith, isn't it?"

"Yes," he said with a smile, "my name is Smith."

"Why," I exclaimed, "you're Tyson!"

"How do you know?" he inquired quickly. "Someone on board told you, I suppose."

"No," I said. "I knew you the moment I saw you—met you years ago."

"Oh, then," laughing, "you've been having the loan of me."

"No," I replied, "though I would not mind if I had the loan of half your fortune for a few years."

"You'd find it was a great weight and a great responsibility."

"I dare say—but I'm willing to take my share of the responsibility if you'll only share the money."

"Now, what would you do with it?"

"Share some with some people whom I know, and enjoy life with the remainder."

"What do you call enjoying life?"

"Well, I'd buy a yacht, and travel and see all that was worth seeing outside Australia."

"That's not a bad idea," he replied thoughtfully; "I should like to travel, too—see America especially—if I could only get away without anyone knowing it;

but I can't. I couldn't leave my business."

"Why not? Get someone else to manage it."

"No one could."

"You'll die someday, and then——"

"Then it won't trouble me. But now it's my life, my pleasure. I could not live without it."

"But why not sell out and travel? You'd find plenty to occupy your mind and amuse you too."

"Nonsense! Who could buy me out? I could sacrifice, but not sell; and I'm not going to do that, though I should greatly like to go to America. I had an invitation from an American millionaire, who asked me to go over to 'Frisco, and he would show me all over the Western States; and then I could go on to his brother, who would show me all that was worth seeing in the East; but I couldn't get away."

Then we sat down on the grating behind the wheel and had a long talk; not only then but often afterwards while he remained on board. Our principal subjects were stock and stations.

But he gave me a sketch of his life just to show me how it was possible for a man to get rich in Australia, no matter how poor he might be, if he only bent his mind to it, and learned to save his money as he earned it, instead of spending it on drink or amusement.

But I must say here (as I then said to Mr. Tyson), according to his own showing, a man could not make money now as he did then. It was the times in which he lived that made him, or at any rate gave him the chance of making a fortune. He seized it; was capable of making the most of the opportunity, and using his money to the best advantage; denying himself almost every pleasure, and devoting time, energy, and brains to the business of amassing wealth, while others who made money quickly spent it as though there never again would be such a thing as poverty in Australia.

Mr. Tyson once said to me: "When a young man I was working for 10s. per week, and I had to work hard, too. I went to look after a lot of cattle that had just been bought, and it took months and months of hard riding and constant watching before I could get them to settle down. I lived in a humpy by myself, and, as the blacks were not to be trusted, many a night I had to leave its shelter, poor as it was, and camp out where they could not find me. Often of an evening, after being in the saddle all day long, I would, as I rode home, strike the fresh tracks of a mob of cattle that had crossed the boundary since I passed there in the morning, and there was nothing for it but to get back to my camp, catch a fresh horse, roll up a blanket, shove a handful of wheat in my pocket if I had none ground—for at that time a man had to grind his own wheat if he wanted flour, so I used to carry a handful or so of the raw material— and chew it as I rode back to where I had seen the fresh tracks, and camp there all night. If in luck I might get a 'possum and roast him on the coals for supper; if not I ate the wheat, lay down till daylight and then followed the tracks till I found the mob and brought them back. Perhaps it would take me all day long, and bar a mouthful of wheat and a drink of water I would have nothing to eat till I returned to my hut at night. And all this for 10s. a week. Yet I managed to save money.

"I remained at this work for about twelve months, when the run was bought, and the owners brought down some breeding cattle and let them go with the bullocks that gave me such a lot of trouble, and, to my surprise, a slow, lazy old man was sent to take my place. I told the boss that the old fellow could never manage to hold the cattle on the run, as it had taken me all my time, night and day, to prevent the bullocks from getting away; and now that there were a lot more it would require two smart, painstaking men to look after them properly. I thought they would soon find they could not do without me, because not only had I done the work of two men but I knew the country thoroughly; and not one of them did, least of all the lazy old man. But they cleared out and left this old fellow to mind the whole herd. A nice mess he will make of it, I thought. They will be sorry they didn't keep me when they find the cattle have strayed all over the country. However, I passed that way again in about three months, found the old fellow pottering about, and taking the world easy as usual; and to my astonishment found also that the cattle were alright, though I was not there to mind them.

"Here I was taught—and learned by heart— a lesson I have never forgotten, which is— that no matter how good a man is, or what his position in life is, he can be done without. I always remember this. When you said to me, 'You will die someday, and then someone else will take your place,' this lesson that I had learned so many years ago was in my mind; I know I can be done without. The world will go on just the same when I am dead. But while I'm alive I intend to manage my own business. I could not live in idleness."

"Well, but how did you make a start on the road to riches?"

"Oh," he said, "though, as I told you, I only had ten bob a week, I managed in time to save a hundred pounds, and bought or took up a bit of a run for £10"—I think the narrator said it was Yanko station[11], which I know is now worth over £100,000. "At that time, you could buy young cattle for 10s. a head, and as there were any number of cleanskins a man with a little capital and plenty of energy could soon put a bit of a herd together. My brother and I worked together on the station. Sometimes one of us went away to work for wages. Then, when we had a few fats fit for market, I took them down, and I tell you I had to be very careful to clear £20 out of my trip, as you could only get about 30s. or £2 for a fat bullock at that time. Talk of droving! That was when you had to rough it. No dray and horses, no travelling bag, and no pyjamas. No, I used to take them down by myself; only one horse, which I led most of the time, as

---

[11] James never had an interest in Yanko Station. He paid £10 squatting licence for Bundoolah on the Billabong Creek: Zita Denholm, T.Y.S.O.N. Triple D Books Wagga Wagga 2000, P 38. But soon after moved to Toorong as his first paying property

be had to carry my blanket and saddlebags with my rations in. At last gold was discovered, a big rush followed, and fat cattle were sold at high figures. Bullocks that a year before I could only obtain £2 for now brought me in £14, £16, and even £18.

"Then, Mr. Tyson," I said, "it was the times made you?"

"The diggings," he replied, "certainly gave a start to everything; other men had the same chances that I had; why did they not become rich? No, I flatter myself I'm the only millionaire in the world who has made his money through sheer business. Lots of men in America have made larger fortunes than mine, and much more quickly than I have, but it was by speculation—lose or win. I never go in for that; every penny that I make is made in the way of business, slowly but surely. Once you reach a certain thing, then money begins to accumulate of itself."

"But of what use is such a vast amount to you? We were speaking a while ago of a Mr. P., whom I know you helped; but you say you are afraid you will have to let him go, as he is too deeply in debt. Now, Mr. Tyson, don't you think it would give you much more pleasure if you were to stick to him and give him another start than to know that the £20,000 or £30,000 it cost you was lying idle in the bank?"

"I never let my money lie idle," he replied, "and as to sticking to P, that's not business at all, and I confine myself strictly to business in my relations with my fellow-man."

"Women, they say, you hate, and that you won't employ married men?"

"On the contrary, most of my managers are married, and I believe in it—for them. It keeps them from fooling about the country after girls; and once a man is married his wife will take care that he stays at home—which is all the better for me."

"Why do you always travel under the name of Smith?" I asked.

"Because I like to go through the world quietly, never making a ripple on the surface" —a favourite phrase of his. "I don't want anyone to know my business, where I am, where I'm going to, or where I came from. The manager of the station I'm going to look at now does not know I'm coming. When I leave there, he will not know where I'm going to, or at what time to expect me again. Thus, I go through life quietly, never making a ripple on the surface."

Then we discussed life in the bush, and he told me of some of the hardships he had gone through. Once he was coaching it, and it rained all day, and when night set in the country was covered with water. The horses were unable to draw the empty coach. So, the driver said the only thing they could do was to take them out, leave the harness in the coach, ride two and lead two, to the next station, about five miles from where the coach was stuck. They started, the horses plunging through the water, and bogging nearly to their knees at every step. After riding for about an hour, he found that he had lost his guide, the driver; and it was so dark and raining so heavily that he could not tell whether he was on the road or not; so, he went floundering on for about another hour, when the horses came to a dead halt and refused to be urged forward. In pulling them about and kicking them to make them go his toe struck against a wire—he felt it with his foot, and found he was against a wire fence, and, not knowing which way to turn, dismounted in about 3ft. of water, climbed on top of a post, and resting his feet on the wires remained there all night long, hungry, cold, wet, and miserable, only to find when day broke that the whole country was under water. And the station which he had so fervently hoped to reach was —just inside the fence. "Yes," he said, "there it was within a hundred yards of me; the horses had brought me straight, but the darkness was so dense that I could not see the house, and so spent about as bad a night as ever a man put in, sitting there on that wire fence —a regular nightmare of a night, which I have never forgotten."

"Oh," I said, "we spend many a wet night riding around the bullocks when on the road. Of course, I'm forced to do it, so don't think much about it, because I'm poor; but if I was a millionaire like you, I'd take precious good care not to get caught like that."

"What would you do?" he asked.

"Why," I replied, "I wouldn't have gone at all; I'd pay someone else to do my business there."

"But suppose no one else could do your business and you had to go?"

"Then, having more money than I could possibly spend in a lifetime, I would not consider the cost, so I'd buy a balloon, run a railway out there, or if I could do neither of these I'd write to Cobb and Co., and they would have four fresh horses to meet me every four miles, and then one could slide along as smoothly as a sunbeam steals through the cracks in an old slab hut."

"Ah," he said, "if you go in for that sort of thing, you'll never be rich."

"No," I re-joined, "I would not think of doing it now; no more than you would years ago when you were stock-riding for ten bob a week. But if I had a few million to my credit I certainly would try to make travelling easier, and guard against camping on a wire fence all night in the wet, or begging a quart full of water from a swagman in a drought. What's the good of money if you don't enjoy it.?"

"Different people have got different ideas of enjoyment. Spending money is not mine. But I should like to go up with you and have a trip in with those bullocks. Nothing I'd like better."

"Well," I said, "come along; I'll give you thirty-five bob a week, the same as the other men; or, as you say money is such a heavy responsibility, you can take my billet. I'll gladly take yours with all its care and responsibility."

"There's one thing," he said, "if I did take your billet, I would not knock cattle about by stringing them along as you fellows do nowadays when you want to count them. I'd just ride quietly through them while they were spread out feeding and count them in little lots—thirty here, fifty there, and so on."

"But," I said, "I don't think you could do that with a big mob. Of course, 'twas easy enough to count those little lots of fats that way, but a big mob take some time to count, and they are constantly shifting their positions; you would be very likely to make a mistake." But he assured me he had often counted cattle correctly like that.

Thus, we spent most of the day yarning, as "a fellow feeling makes us wondrous kind"; and as I was a "cattle man," and he would "talk cattle" from morning till evening, and as he did not seem to know anyone on board, we were constantly together, discussing stock and stations principally, but touching on all kinds of things. For instance, he informed me he could trace his descent straight back to a Saxon family, who owned the land at the mouth of the Humber, before the Norman conquest, and that he could claim a title if he cared to; but he preferred going through the world quietly, and 'never making a ripple on the surface.'

Just before leaving the vessel, he came up to me with the old valise under his arm and said, "Good-bye, Mr."

"D____y," I 'said, "is my name. Don't forget it when you are making your will." He laughed, shook hands, and, promising to remember, stepped on board the tender. That was the last I saw of my rich fellow-traveller, and I still remain poor.

The following may not be of interest to all readers but some readers may be interested in descriptions of Tupra when James held it.

Riverine Grazier, Friday, 9 March 1894, P 2.

SOME RIVERINE FREEHOLDS.
"Tupra,"
[By Harold M. Mackenzie.]

The well-known name of Tyson is one to conjure with. One immediately associates it with vast territorial possessions that would not disgrace the Czar of Russia himself ; with millions of sheep and bullocks and horses that if driven in single file might possibly put a girdle round the earth; with a man who neither smokes nor drinks, and, strangely too, is unmarried ; with a muscular man : with a man possessed of wonderful energy and a capacity for seeing further through a brick wall than most people ; in a word with a millionaire, self-made, and without vices. I speak from report which in the main I have every reason to believe is correct of the Hon. Jas. Tyson. That there is no man living without faults is an accepted truth, not to know which in oneself is a calamity, but knowing which and still not trying to correct them is a heinous offence. But what has moralising to do with an article on station matters, though in truth these slight digressions from subjects I am so accustomed to are in a measure a species of relaxation and pleasant diversion from the groove in which my mind is continually running. Leagues upon leagues of salt-bush plains, varied with nothing more interesting than a dry swamp, rabbits, and alas! "Dead marines," are not calculated to excite the imagination unduly. Still there is much food for reflection even in empty beer bottles, and the possibility of their contents having struck imaginative chords in the consumers.

But I am arrived at Tupra, having journeyed from Till Till by way of Oxley station, where, through the kindness of Mr Macdonald, I made a short visit on the road hence. Seven hundred thousand acres, even when said quickly, is a large area and enough to make one take more than ordinary interest in a man who for years has had the use of them for running stock. Ancient and modern Tupra, however, are two very different places when one compares the original area abovementioned with that of 435,000 acres representing the area at the present time? In the latter amount are included 100,000 acres freehold, and the balance 335,000 represent the Crown leasehold portions, upon which something over 2d per acre is the annual rent paid, or in other words about £2800 — a substantial sum of money in this age of mortgages stock tax and rabbits. With the former, however, Mr Jas. Tyson is fortunately unacquainted, in so far as being a borrower goes, which naturally leaves him a generous hand to make whatever improvements may be deemed necessary for the good of his many stations. It may be truthfully asserted that a man, owning at least a dozen different stations in the various colonies, owing nothing on any of them, content to live on the goods the gods provide, luxuries being regarded as superfluous, stands forth in the pages of pastoral history as a character absolutely unique. With the decrease of acreage at Tupra, a corresponding falling off is naturally to be expected in the stock, which of late numbering about 140,000 head, are today reckoned at 80,000. In addition to the resumed area being taken up, the ever present and rapidly propagating rabbit plague has done much in reducing Tupra's carrying capacity, which, as things stand to-day, may be set down, if overstocking would be eschewed, at one sheep to seven acres. That Mr Jas. Tyson is cautious, maybe over cautious, in this matter is a thing that cannot be too highly commended, seeing that rabbits, as things are shaping, promise to be a worse scourge in the immediate future than ever before. Though not a pessimist by any means, the outlook, judging by the marvellous increase of the vermin, if not reduced this summer, promises a lamentable picture in prospect.

For the last 40 years Tupra has been held in the name of Tyson, that is to say, Messrs James and Peter entered into partnership for this as well as other estates on the Lachlan, but a dissolution having taken place this run is exclusively owned at the present time by the former gentleman. Adjoining Tupra we have Corrong (executors of the late Peter Tyson) lying east by north, and on the west blocks taken in by homestead lessees. The Lachlan offers a magnificent water frontage of no less than 28 miles, whilst the Murrumbidgee does duty in a similar manner on portions of the southern area for a distance of something like six miles. The paddocks, with two exceptions, are all subdivided, so that each has natural water for stock purposes, furnished either by means of rivers, lakes, or canals. Where expense is not a matter of grave moment, providing the back country with water by means of aqueducts, is of course the only true and efficient means of ensuring a constant supply. This scheme Mr Tyson has entered into with a spirit at once enterprising and lavish, so that tanks and wells are quite unnecessary commodities on Tupra Lakes years ago, which were but mere apologies for the name, now contain, in some instances, as much as 30ft. of water, fed by canals running from the Lachlan. Two of these schemes, a canal running 10 miles in length from the river into Lake Duckshot, whilst another constructed from the Lachlan into Box Creek, represent a figure, taken in the aggregate, of no less than £12,000. Of course, schemes of this kind are, practically speaking, infinitely better than tanks which are constantly silting up, and after five years become useless, or sinking wells, the water from which, on this run at least, is most unpalatable even for stock. Tupra may be described as a lake country, the abundance of natural water on same being furnished by means of canals, or else by the overflow of the Lachlan, which always ensures enough, and more than enough, even in the driest seasons. The only other place which I can recall to mind where schemes on the same principle have been boldly carried out by the owners is Toogimbie, whose canals are after the pattern of Mildura.

Roughly speaking, the Tupra paddocks may easy be said to afford subsistence for 10,000 sheep, whose conditions at the present time, considering the trials and privations which they undergo in this district, is exceptionally good. The management of this station is now in the hands of Mr Patrick Gleeson, successor to Mr James Tyson, jun., who came to this district in '63 and to whom for his voluntary information in connection with Tupra I am necessarily indebted. When the partnership existing between Mr James Tyson and Mr Peter Tyson was in force, the type of sheep bred then at Tupra had a distinct dash of Mungadal blood throughout them, supplemented with several experimental crosses that for obvious reasons were, not the signal successes at one time predicted. It was more, however, on account of Mr Peter Tyson that these experiments were indulged in rather than that of Mr James, whose policy in the matter of sheep breading is, stick to what you have without interchange of blood. In reference to the crosses, rams of the Negretti breed were once introduced, the wool of which is remarkable for its very dense and short character, being combing, not clothing wool, and with a staple that in the end was calculated to do irreparable injury to flocks, as a body. This breed being unanimously tabooed by the squatters hereabouts, the next and uppermost thought in their minds was to restore once again the clear staple or Ercildome (sic) type of wool, which eventually took precedence over the German cross. Another experiment tried, and found wanting, was the intermixture of Tupra blood with that of the Southdown, but what result was expected by such a cross was not very clear at the time. The demand for frozen mutton in Britain was not then in existence, the first and all-important question in a pastoralist's mind being fleece, not carcase. That the progeny of Southdown ruins and Tupra ewes was all that could be desired from a butcher's point of view cannot be disputed, wethers, for example, turning the scale at 80 pounds. But wool was quite another matter, and great was the falling off in this respect, the fleeces being of a light, fuzzy, and blanketly character. The Southdowns, as a consequence, went by the board. Mr James Tyson, seeing the fallacy of further experiments, adopted, as soon as the dissolution of partnership took place, a definite and rigorous course of sheep breeding, which he has not deviated from since that time. The rams purchased from Dr. Laing, of Mungadal, bearing no comparison of course with those of the present day, were nevertheless the best that could be procured in the district, their robustness of wool, then as now, being a distinguishing feature. From these sires the Tupra flocks have continued their descent, a hardy, hunger enduring race, that are unmatched in respect of chewing imaginary grasses. That Mr Tyson is a man who endeavours to feed his sheep, no matter the season, no matter the rabbits, no matter, in fact, what occurs, is a characteristic that all on Tupra must credit him with. "Make your improvements," he says to Mr Gleeson, "kill the rabbits, keep the sheep alive, and never mind the expense." When a manager has carte blanche to go ahead, and do what he likes without having to consult his employer, things should wear at least a hopeful aspect, and this may be said to be Tupra's position to-day. If not showing great results in the matter of returns, the place is at least resting on its oars until the arrival of better times. May they soon return! Of the 74,000 sheep shorn in 1893, together with 20,000 lambs, it was not to be expected that with a rainfall for the preceding twelve months of 7inches only, that anything startling in the way of weighty wool would result. Rabbits and want of rain had told severely on the country, so much so that the salt-bush in many parts of the country was completely and effectually razed from the face of the earth, sheep coming in to be shorn semi- starved and wholly pitiable in appearance. The country may have been pushed to its utmost carrying capacity, but the chief cause at work was the rabbits, who were literally eating the place out. Despite the untoward circumstances of last winter, the average weight of ram's fleeces was between 8lbs. and 9lbs. and that of other sorts showed a return of 6 pounds. Light as these may appear it is something to know that the sheep were all shorn, and no losses beyond the usual percentage have to be chronicled. The rainfall, which was so abundant out north, has made no appreciable difference to Tupra as far as grass is concerned, though no apprehension is felt on Mr Gleeson's part should a second visitation not occur between this and early April. The quality of Tupra wool, owing to a dry season, has not had a fair chance of doing justice to itself at the recent London sales, being characterised as a good quality wool, but earthy, the price realizing as per cablegram, 6¾d. What wool indeed would stand the test of a connoisseur's eye when stock is pushed to extremities such as were experienced in this district last winter.

Like many other flock-owners, Mr James Tyson is not a believer by any means in travelling sheep to the mountains when a dry season sets in, more especially in these days, when the expense entailed, would be out of all proportion with results, His policy, short, sharp and summary though it be, is reasonable withal. "If they can't be sold," he says, "they must die." Fortunately, there has been no occasion to witness anything so deplorable, since surplus stock can always be knocked on the head at Hay or Moama at prices, all the same, which are enough to make by-gone squatters turn in their graves. A draft of aged ewes was recently despatched to Moama, bringing 2s per head, something after all as prices go now-a-days, but as stock must be got rid of at all hazards better to accept starvation prices than none at all. Early lambing is a desirable thing in Mr Gleeson's opinion, giving the young stock time to mature sufficiently before being shorn, thereby enabling them to withstand the operation better, and in the event of becoming forcibly weaned, for it is well known that many lambs lose their mothers at shearing, they are more capable of looking after themselves.

Touching upon the general appearance of Tupra as a grazing property, the natural grasses which thrive in a good season are not unlike what one is accustomed to see on the Murrumbidgee, more especially on the western portion, which is best for stock in summer, owing to its swampy nature, and rich, succulent herbage. The country here, abouts is looking good enough at the present time, and will carry the sheep well until such time as it is deemed advisable to run them into the frontage paddocks. An immense quantity of silver grass is noticeable everywhere, having been beaten down in places by the recent rainfall, innutritious and worthless as far as feed goes. In many places, the rich river bends more especially, stock have either eaten the grass out, or it has become so dry that it serves no practical purpose. It is somewhat surprising to hear that no less than a hundred rabbiters are engaged at Tupra, receiving £1 per 1000 for scalps, the men being allowed to retain the skins. The rabbiters have been making a good living at the work, destroying them wholesale with dogs, traps, arsenic water, phosphorated wheat and other means, though the campaign has received a temporary check since the rain. In order to convey some idea of the stupendous result of a year's warfare, no less than 1,700,000 head were slaughtered, and yet in spite of all, these prolific vermin promise to be as numerous as ever. Besides these hundred men, three poison carts are always kept going, wages being paid to the men at the rate of 15s per week, so that taken altogether it is quite within the mark to state that £2000 is annually expended on rabbit extermination. Referring to the question of reduction in wages, which, as a rule, are now 15s per week instead of £1, it seems that the men have deemed it wise to accept same with a good grace, knowing that times are not what they were, and that plenty are ready to fill their places should they wish to vacate. It may be mentioned, however, that Mr Gleeson has been instructed by Mr Tyson that whenever occasion demands a return to the old order of things, he has permission to make the change on his (Mr Tyson's) behalf.

In addition to the number of sheep carried there are also about 2000 head of cattle running in the river bends, 400 of which are in prime condition and ready for market, but which I'm afraid must stop where they are if nothing better than ruling Melbourne prices, minus stock

tax, are to be depended upon.
Tupra, February 13th, 1894.

Melbourne Punch, **Thursday, 26 Dec 1895, P 2.**

PEOPLE WE KNOW

Millionaire Squatter Tyson never drinks, never smokes, never swears, and is a bachelor. As Disraeli said of Gladstone, 'He has not one redeeming vice.' Added to this he is very frugal and unostentatious. Upon his appointment to the Queensland Legislative Council, he was travelling down to Brisbane to take his seat, and had a cheeky jackeroo for a companion on the train Journey. 'I suppose you're taking down a dress suit, Mr. Tyson?' said the youth. 'No,' replied the squatter, 'only a clean pair of moleskins. Dress clothes have had their innings; it's, time honest moleskins had a show.'

The dandy jackeroo subsided."

CHAPTER 4

Articles from his death to the trial to determine his domicile in November 1901

On Sunday December 4, 1899, James died. There was a plethora of tributes of which a very small samples are presented in this chapter. They require no commentary except to say that it was sad that he died a little more than two years before the Federation of Australia he so strongly advocated.

Brisbane Courier, **Monday, 5 December 1898, P 4.**
DEATH OF THE HON. JAMES TYSON
FOUND DEAD IN BED
(By Telegraph from Our Correspondent)
PITTSWORTH, December 4.

The Hon. James Tyson, M.L.C., was found dead in his bed this morning at Felton station.

On the manager of the station, Mr. Buchanan, going in to breakfast, he was surprised to find that Mr. Tyson was not at his usual place at the table, and he instructed a servant to go and knock at the bedroom door to try and arouse him. This the servant did, but, receiving no reply, he informed Mr. Buchanan, who then went himself and opened the bedroom door, only to find Mr. Tyson lying dead in bed. He seemed to have died without a struggle. The deceased had been ailing for the past fortnight, but, notwithstanding all advice, he absolutely declined to see a doctor. His usual time for rising was 4 a.m., but he intimated to the servant previous to retiring on Saturday night that he might possibly not rise so early this morning. His nephew, Mr. Tyson Doneley, has been telegraphed for, and pending his arrival the body lies at Felton station.

ARRANGEMENTS FOR THE FUNERAL

Mr. T. Macdonald-Peterson, who was summoned to Felton yesterday by a telegram from the manager, Mr. J. Buchanan, telegraphed last night from Toowoomba:-" It is arranged that the funeral of the late Hon. James Tyson will leave St. James Church, Toowoomba, on Tuesday afternoon, at 2 o'clock, for the Toowoomba Cemetery. This will be ninety minutes after the arrival of the morning train from Brisbane.

The flag flying half-mast high on the Parliamentary buildings yesterday afternoon notified to the public the passing away of one of the numbers, of the Legislature, but very few people anticipated the demise of the Hon. James Tyson, M.L.C. Not many days ago his tall and well-known figure was seen in our thoroughfares, and he was transacting business with the forethought and precision for which he was remarkable. He was not a man who entertained his acquaintances with recitals of bodily ailments, and he somewhat resented personal inquiries upon that subject; but it was apparent to all his friends that his efforts to keep pace with the business of the day taxed his strength to the utmost, and that physical weakness was dominated by sheer strength of will. Upon the return of the Hon. James Tyson to Felton his condition did not improve, and, as already reported, he was found dead in his bed yesterday morning, having evidently passed away in his sleep.

For more than twenty-five years (writes a contributor) the Hon. James Tyson has been a unique and prominent figure in the history of Queensland, and a powerful factor in the development of its resources. He is a man who has left his mark upon the land, and not only upon the land, but upon everyone with whom he came in contact. He possessed strong personality, verging perhaps upon eccentricity, and original and lofty conceptions upon the duty of man, the absolute power of the Creator, and the ordering of the great events of life. In the same way that he carried his speculations into the business transactions of life, so his speculative mind sought to penetrate the dark shadows of a future world, and in the book of Nature, under the stars of heaven, he worked out his theories of futurity. To those intimates who had the privilege of meeting this prince of squatters on his own station residence at Felton it was a matter of astonishment that such a practical man should take up so much that was abstruse, and collect in his library so many volumes upon the great problems of life. These were topics that he was never tired of discussing, and his range of reading took in most of the modern works of the day. But as a politician he did not care to shine; for there was much of the roving patriarch in his nature-his flocks and his herds and his large territorial possessions were the glories of his life, and the objects of his solicitude. His untold wealth was a matter he seldom discussed. When the subject was broached, he often said, "I am happiest under the stars of heaven, with a bluey for my pillow, and a billy of tea by my side." As an employer who himself had once been the employed, he was exacting, upright, and severely just. On the Darling Downs, where he was so well known, the Hon. James Tyson commanded the respect of all his men. He was often a silent benefactor, and an unostentatious friend.

Sydney Morning Herald, **Monday, 5 December 1898. P 4**

DEATH OF THE HON. JAMES TYSON, M.L.C.
BRISBANE, Sunday

The death of Mr. James Tyson removes one of Australia's most prominent settlers. To the general public the name is known as that of one of our squatting kings, who has been most successful in acquiring wealth from the broad acres which he added from decade to decade to his estate. Beginning life as a poor man, he brought to the pursuit of his business an untiring industry and native sagacity, which were aided by such unbounded resources as the continent possesses. In these he had as great confidence as he had in himself as their exploiter, and he could be very severe on any who doubted them. Beginning life with nothing, he developed qualities which, while contributing to his success, were perhaps generally regarded as eccentricities that remained with him, as they commonly do in such cases, to the close of life. He was the determined enemy of everything in the shape of luxurious living or extravagance in expenditure on anything that did not contribute to the acquisition of material wealth. And he traced the bulk of their failures of those engaged in his own line of business to the neglect of the simplicity of life which was so marked a feature in his own career. Nevertheless Mr. Tyson possessed other qualities of a different character. He was not destitute of sympathy for deserving objects, and under the advice of those in whom he trusted, and where he was sure that his benefactions would not be misplaced, he contributed large sums to charities or public objects in which he took an interest. Altogether Mr. Tyson was a remarkable man-made so to some extent by the isolation of his life and the absence of those family associations which usually modify peculiarities. He has done his own work in opening up and developing large tracts of country, from which a vast mass of wealth has been acquired, and distributed and redistributed almost entirely in employment to colonists or in public revenue to the Government.

Argus, Tuesday, 6 Dec 1898, P 6.

JAMES TYSON, MILLIONAIRE
A Remarkable Career
(From the Argus Correspondent.)
Brisbane, Monday.

Mr. Tyson had never been so ill as to require the services of a medical man. On this occasion he refused to see one, though pressed to do so. On Saturday night he felt no worse than he had done for a few days.

He was one of a family of six sons and four daughters. His brothers, John, Peter, William, Charles, and Thomas, and his sisters, Mesdames Doneley, Heron, Hewitt, and Moore, all predeceased him. He leaves one step sister, Mrs. Shields, of Wilton, near Sydney.

Mr. Tyson's life was little understood by Australians generally. It is known that he was charitable in an unostentatious way, and at Toowoomba yesterday Father Fouhy referred in touching terms to the virtues of the deceased, eulogising him for his financial help to St. Patrick's Church and to the Christian Brothers' School. He was not a man who entertained acquaintances with recitals of his bodily ailments, and somewhat resented personal inquiries upon that subject, but it was apparent to all his friends that his efforts to keep pace with the business of the day taxed his strength to the utmost, and that his physical weakness was overcome by sheer strength of will. It was apparent, however, of late that he was growing old. His tall spare form in recent years lost its uprightness, and his carriage its characteristic grace. He became deaf, a point upon which he was very sensitive, and his voice lost its power.

To those intimate friends who had the privilege of meeting Mr. Tyson at his station residence, at Felton, it was a matter of astonishment that such a practical man should take an interest in abstruse subjects; and should have collected in his library so many volumes upon the great problems of life. These were topics he was never tired of discussing, and his range of reading took in most of the modern works of the day. As a politician he did not care to shine. His flocks and herds and his large territorial possessions were the chief objects of his solicitude. His immense wealth was a subject he seldom discussed. When it was broached, he often said, "I am happiest under the stars of heaven, with a bluey for my pillow and a billy of tea by my side.

Mr. Tyson was not a polished man, and cared nothing for social distinction, but he was not without family pride, and traced his pedigree to a Saxon ancestor, who owned property at the mouth of the Humber, before the Norman Conquest, and came of a good Cumberland stock, yet he was not a vulgar man. His riches were not displayed. His economical methods and ascetic habits prevented this. He had one peculiarity, namely, addressing every man as 'Mister,' "What did you say, Mister," being a frequent query, with his increasing deafness. He was very fond of children. He never touched intoxicating liquor nor tobacco, and no one remembers having heard a profane word from his lips. The purity and simplicity of his life disarmed those inclined to covet his riches. Personal indulgences he knew not himself, and would not tolerate in others. Out on his western runs his tall, spare form— he was 6ft. 3½in. — was not unwelcome, though he was seldom complimentary, and his sense of justice was very strict. In political life he took little part, but for a few sessions he attended the sittings of the Legislative Council pretty regularly. The only speech he ever made was on the Marsupials

Bill, when pressed by his friends to give the House the benefit of his experience. His speech was a disquisition on the nesting habits of kangaroo rats and paddymelons, but Hansard misplaced the names, which so disgusted him that he never addressed the House again. "It was my only speech, and they spoiled it," he said.

Mr. Tyson was never married. Here is a quotation from an interview with him some time ago, "Women, they say you hate, and that you won't employ married men?" "On the contrary, most of my managers are married, and I believe in it for them. It keeps them from following about the country after girls. Once a man is married his wife will take care that he stays at home, which is all the better for me." He usually travelled under the name of Smith, and when asked why he did so said, "Because I like to go through the world quietly, never making a ripple on the surface." A favourite phrase of his was, "I don't want anyone to know my business, where I am, where I am going to, or where I came from. The manager of the station I am going to look at now does not know that I am coming. When I am there, he will not know where I am going to, or at what time to expect me again."

Reading Mercury, UK, Saturday, 11 Mar 1899, P 10.

SOME RECOLLECTIONS OF JAS. TYSON
A Native-Born Australian Millionaire and Squatter
BY A PIONEER (Mr JOHN PHILLIPS)

The death of James Tyson, "a squatter King and Australian Millionaire," at Brisbane, Queensland, on 3rd December, 1898, recalls to my recollection his earlier career and subsequent knowledge for upwards of thirty years of his life, which is certainly a remarkable one, showing how a self-made and self-educated man may attain a great wealth and position from a humble and small beginning, through sheer persistency of purpose, "steadiness" and great abstemiousness together with cute perception and strong common sense. Though in all his vast transactions, punctilious and exacting to the letter, yet withal straightforward and honest in all his dealings, never losing a chance in a bargain, and always prepared to buy in the cheapest market.

His first appearance in the district in which I resided was, I think, 1846 or 1847, and he was then occupying some country on the lower junction of the Lachlan with the Murrumbidgee River in New South Wales, having travelled down with a small herd of dairy cattle and started a dairy station. His appearance, then a young man of about 25, was in his favour—tall and wiry looking, black hair and beard, with small, dark beady eyes, with the shrewdness and self-confidence peculiar to the native born Australian. During his first years of dairying no money could be made, as there was no market for the produce, and so he set to work to make cheese, and once in the year used to take his cheese in a horse and cart to the port of Melbourne, and generally spend the night at our station, it being on his way, when he used to present us with a cheese, which was more like a cannonball, and as hard, taking a tomahawk to break it. On one occasion he called on his way travelling up to Sydney with a led horse. At the time the rivers were in flood, with no means of crossing. On asking him how he intended crossing the Murrumbidgee he said there was a punt lately put on, some 200 miles higher up the river but as he had no money to pay for puntage of his horses we managed to raise him five shillings. On meeting him again some six months afterwards, and asking how he managed, his reply was, "I was not going to fool away that five shillings you lent me. I just faced the horses into the river, laid hold of their tails, and swam across." So that he took that money up to Campbell Town, travelling some 500 miles with it in his pocket. Such was the man.

When gold was discovered in 1852, he gave his attention to the purchasing of sheep and cattle and droving for his two brothers, William and Peter, who started in Bendigo as dealers and butchers, keeping slaughtering yards, supplying them with beef and mutton, which they retailed at considerable profit to other butchers and to diggers. This business was carried on to their mutual advantage for some three to four years when the brothers separated. It was then that James Tyson went into purchase on his own account, buying stations with such good judgement, having sufficient stock on them at the time as would nearly clear the purchase money; but never sitting down, always moving from one place to another, on the lookout for bargains. No one had a quicker eye or shrewder guess as to the fattening capabilities of the runs he passed through, or of the probable value of the stock on it. He always had good horses under or carrying him, for in those days the roads were not as they are now, and his usual mode of travelling was with a led pack horse, on which was carried his blankets and provisions, quart pot etc., for he generally preferred camping out, where the grass was good for his horse, to putting up at strange stations, especially should there be ladies there, preferring to ride up in the early mornings, perhaps partake of breakfast, and talk over prices of stock and markets. He often told me in travelling he only needed a little salt in his pocket to flavour anything he might get on the road, a piece of soap to wash his shirt with when needed, and from his partiality to green stuff, and the way he lived and travelled in those days various names were given him such as, "Hungry Tyson," "Nebuchadnezzar," and "Poor Man Tyson." In one of those interesting books by Rolf Boldrewood. "The Colonial Reformer," Tyson introduced as Mr Abstemious Levison, being camped in the middle of the day, meets the young Reformer, and one can almost imagine hearing his words: "One never is lacking a luncheon here mister, plenty of good salad and wholesome grasses all round; only needs a pinch of salt."

On one occasion, travelling down the country in his company in one of Cobb and Co's stage coaches, we came at midday where we change horses, when I said, "Come on, Tyson, we can get a basin of soup here." His reply, "No mister," which was his usual mode of addressing anyone, "I am a poor man, and can't afford to spend two shillings on a basin of soup, and then not have time to take it. When I travel, two-pennyworth of biscuits and a bottle of lemonade will do me until sundown." Lemonade was, I believe, his only weakness. His two first purchase of stations, which proved remunerative to him, were the north Deniliquin Station, in the Riverine district of New South Wales, purchased from the Royal Bank of Australia, and originally taken up by Ben Boyd of the Wanderer Yacht notoriety. The other was Heyfield Station of Gipps Land, (sic) Victoria. On the former he built a brick dwelling-house—the first of its kind in our district. On its completion, and being our nearest neighbour, I remarked he should give a house-warming, providing champagne and cake, and invite a few friends from the township of Deniliquin to partake, and inspect his new house. He then, fixing the day, left it to me to ask whom I thought would care to come. Consequently, I invited five or six and amongst them two ladies who were curious to see Mr Tyson and his new house.

On the day fixed we were riding up in a party on horseback. Tyson was looking up the road, watching for our arrival, and happened to notice the ladies' skirts flying in the wind, on which he at once retreated through his house, crossing the river at its back, and made straight for the bush on the other side. Consequently, there being no host on arrival, I had to do the honours at the reception, etc. On meeting him again a few days afterwards, and enquiring his reason for so abrupt departure, he said, "You know, Mister, I don't care for lady's society, and you should not have brought them." Many reasons have been assigned for his antipathy to females, as he usually termed them, and his generally avoiding their presence was, I believe, a good deal from shyness, he being of a retiring nature. It was his impression that some woman would marry him for his money and squander it away. But when he did happen to be in the presence of a lady his manner and behaviour were most courteous and respectful. He was, moreover, very partial to and kind to children. I always considered him as a man of high moral rectitude, and never to my recollection did I hear a blasphemous or obscene expression emanate from his lips.

It was after he had disposed of his station on North Deniliquin, and left our district, that he entered on larger purchases and transactions in land and stations, in all three Colonies, buying up land from the Government in blocks of thousands of acres, not only in New South Wales and Queensland, but extending his purchases to South Australia and New Zealand; in fact, holding interests in all of the Colonies. I then lost sight of him for a time, though hearing occasionally of his transactions in stock and station; and the last time we met was in 1873.

Having travelled down overland from Sydney to Melbourne, on my way to India, and putting up at the Old Port Phillip Hotel—a well-known resort of squatters in the earlier days—I was wondering if I should meet any old faces that I had formerly known when about 10 o'clock in the evening who should walk into the smoking room but James Tyson. We were mutually surprised and equally glad, I think, to meet again after a span of ten years or so, and after some conversation about the past allusion was made to his increased wealth and prosperity, eliciting from him in reply, "Yes, Mister, any fool can make money in this country easy enough, but my trouble is to know what to do with it when made, and how to invest it safely."

I then proposed that he should lend me £40,000, and instead of going to India, I would proceed up to Queensland and purchase a good, first-class cattle station, then offering, and that I should work it out, he holding the security on same. His reply—"I would write you out a cheque for £50,000 this moment, only I must have security." I then said, "If I had the security to give, I would go to the Bank and get the money." "Well, that is just where it is, Mister."

I recalled to this recollection a quarter of a century past, when I lent him five shillings to pay his puntage over the Murrumbidgee River, and had never asked for interest or security. "No, Mister," he said, "we were always good neighbours and friends," and there we parted, and, in wishing him good bye, I said, "You are returning to Sydney tomorrow, and I am sailing for India, we shall never meet again; fare thee well."

A finer bushman I never knew, nor one who had more thorough knowledge of bush life, and all that pertained to it. He was generous on occasions, and gave liberally when the project was worthy.

Barrier Miner, **Wednesday, 7 Dec 1898, P 7.**

STRAY NOTES.
"Jimmy" Tyson.

Jimmy Tyson stories will for a week or two have a revived popularity. Most of them will probably relate to the reputed millionaire's meanness; and of these from 75 to 95 per cent, will be quite false. Tyson was distinctly not a skinflint. He had somehow got a reputation for detesting ostentation; and he lived up to the reputation – just as other men try to live up to a reputation for open handedness, and pretty often end in the Insolvency Court. He got himself to believe that he liked nothing so well as plain-speaking and he really did like the independence which it sometimes springs from. In such cases men who thought they had made an enemy of Tyson found out that they had instead got hold of a very liberal friend. He wasn't a political capitalist. No National Defence League or association led him by the nose; that was the thing that bushmen liked about him most, and that was why there was very rarely any trouble on his properties. For another thing, he knew how to do the work that every man on every station had to do; and nothing gives more confidence to a good workman than that. People thought he was a boor. The fact was that he was extremely nervous and sensitive – as indeed so many men are who live much in the bush; though the characteristic often conveys an entirely wrong impression to the outsider. The secret of his financial success! Here it is, in words used by himself not many years ago-"I think I know something about stock. I know more than most, anyhow; and I know what I don't know too, and don't go that way. And I've stuck to it."

Argus, **Wednesday, 7 Dec 1898, P 4.**

WEDNESDAY, DECEMBER 7, 1898,

The late James Tyson's fame, which extended to every part of Australia and had its beginnings several decades ago, grew primarily out of his reputedly enormous wealth and was also connected, though in a lesser degree, with the alleged eccentricities. He was both a millionaire and what is commonly termed a character. He had more money than providence has placed in the hands of other persons and he was strikingly unlike his neighbours. As an individual he would have exhibited a certain originality under any circumstances; and the circumstances of a millionaire favoured an extraordinary development of personal quality. No man could have been truer to himself; but that very fidelity isolated him. Nature had so fashioned him that solitariness was inevitable. He had phenomenal success in making money and perhaps a phenomenal gift lay behind the success; but by natural compulsion he was a lonely man and he loved his loneliness. There was kinship between this shy and silent pastoral pioneer and the utmost unpeopled spaces of central Australia in which his life was largely spent. "I am happiest under the stars of heaven, with a bluey for my pillow and a billy of tea by my side." In this simplicity the millionaire whose name was a household word throughout the continent found relief from responsibility.

What his riches amounted to has been the subject of much conjecture, but it is pretty certain that as a possessor he ranked amongst the foremost few. An estimate we print this morning states his fortune at over five million. Like some other members of his class, he is believed to have got more satisfaction out of the making of his money than from the sense of possession. He resented the general curiosity as to his affairs as a general impertinence. He had no comprehension of human nature's interest in millionaires.

But the question of the total of his wealth held only second place in public inquisitiveness. The first question was always. What will he do with it? Altered now to. What has he done with it? Our Brisbane correspondent says Mr Tyson's "flocks and herds and his large territorial possessions were the chief subjects of his solitude" and it was reasonable to suppose that the final disposition of the vast accumulation had occupied his mind a great deal. But he guarded his secret absolutely. None of the small number of persons admitted to the friendship of this strangely-constituted man seemed to know what he had decided to do. It may as well be confessed that in the colony in which his later years were mainly passed there was a widespread hope that the event would prove that he had elaborately thought out some great and worthy benefaction. Our information this morning points to the probability that he died intestate. If this should be confirmed a scramble for the millions by half a hundred relatives will be a strange sequence to a mysterious life. It will suggest that, if his life was not base, it lacked nobility.

Was it to the advantage or disadvantage of Australia that such a man lived his sort of life and built up his vast fortune? We have other millionaires and shall have more; but it may be assumed that there will not be another Tyson. If that is so, is it something on which society may be congratulated? Ought we to be glad that the only Tyson is now dead and gone? Or may we say that this one really nerved his generation? Mr. Tyson lived long because he lived healthily. Nature bestowed on him a splendid physical equipment and a vigorous mind and he adopted habits which, respected and safeguarded them. He was an ascetic in his abstentions. But, if he had no energy for vice or even ordinary pleasure, he had a wonderful capacity for work. He was shrewd and resourceful, and one of the few favoured persons who are well matched with opportunity. And when once he was well started, his money made money in the sense that it often gave its owner a commanding place in competition. Not that he enjoyed immunity from blunders any more than from the disasters which at intervals fell upon the pastoral industry. He made some costly mistakes. But he brought great natural shrewdness and carefully acquired experience to bear upon the many problems which he had to attack. He trusted his own judgment because he gave his whole heart to his business. Of course, there may have been a considerable element of what is commonly called luck in his success, generally there is in such cases. But you cannot eliminate remarkable personality from the success.

It must be confessed that, with many excellent qualities and the fortune of Australia, he was not an enviable man. Nobody wanted to be Tyson; however, many would have liked to get hold of his wealth. He lived in the way that pleased him; he was his own interpreter of duty; but we need not look among the living for imitators of the departed millionaire. He won respect, but it seldom or never kindled into warm admiration and it was not in him to inspire popular affection. But he did not waste money vaingloriously or on the pampering of appetite. Nor did he let it rust. Tyson's millions are distributed amongst the industries of three colonies. Indirectly his fortune has assisted in settling and supporting thousands of families, and, undoubtedly, he took a great individual part in developing the pastoral industry. If he was proud of any personal achievement, it was of the measure of success which had attended his efforts in reclaiming country from its natural unproductiveness. It would be easy to harshly criticise certain phases of his character, and to raise a contemptuous laugh at certain incidents in his life and it must be granted that his strongly-marked individuality was not conspicuously attractive; but, taking him all in all, as the rest of us would like to be taken when the chapter is closed and judgment is prayed, it should be said that millionaire Tyson has left a record which is singularly free of grossness and singularly faithful to his conception of duty. That conception was not the highest; but it is possible to expect over-much from millionaires.

The Queenslander, Saturday, 28 January 1899, P 167.

THE LATE HON. JAMES TYSON
SOME EARLY MEMORIES
By JOHN CAMERON

It is strange what misimpressions take hold of the mind of some men in a community in regard to both its peoples and surroundings. Without the slightest intention to do a wrong to any one, wrong is done, and done to really damage the character of the subject dealt with. I was sitting in the 5.35 train the other evening, bound for Ormiston, when a friend accosted me with the remark: "Old Tyson has gone at last; but he is no loss, for he was only a money-grubber." To this remark I replied: " You make a great mistake with reference to Mr. Tyson, for he was a man of sterling character; a man of more than average ability in other walks of life than the mere accumulation of money; a straight, simple, unostentatious, honest man; a man without guile, who said what came to his mind on any subject that interested him or that came under discussion between him and any one with whom he was debating or arguing a subject. You clearly do not apprehend the high character of the man you condemn." The reply was: "Perhaps not, but that is the general opinion entertained of Mr. Tyson."

Soon after tin was discovered in Stanthorpe - I think it was either 1872 or 1873 the late Hon. James Gibbon and myself wanted to have a look at the tin field. In the same train we travelled by, and in the same first-class carriage, was seated the late Mr. E. R. Drury, and, I think, Mr. R. M. Stewart, and others. Mr. Drury was on his way to Toowoomba, I understood at the time, to make arrangements for opening branches of the Queensland National Bank at that and other towns on the route, but I think at Stanthorpe in particular; and, to judge from appearances when we got to Toowoomba, there had been an arrangement for the late Mr. Tyson to meet Mr. Drury at that place. At any rate, the late Mr. Tyson got into the carriage we occupied, and he had no sooner taken his seat than he said, after surveying the cushions and general surroundings of the carriage, "Well, Mr." (he looked at the late Mr. E. R. Drury but did not name him), " I don't know what my friends will say if they see me in this sort of carriage, for I always travel in the second-class." Mr. Drury and all of us looked at the millionaire with an amused expression on all of our countenances and we waited for further developments. The great capitalist, while speaking had been carefully taking in all the luxury expressed in the soft spring cushions, highly-wrought varnished panelling, &c, when he continued, "You know, I like to be among the working people, for it is from them you have to learn what is going on. I am always learning something from that class of people, but the higher classes read their papers or books in the train, and you hear nothing and learn nothing from them as to what is going on in the country."

When the train got to Clifton, Mr. Tyson remarked, "There are my horses"; and as the train receded from the platform, we could see him carrying his bridle and saddle and wandering into the paddock along the line to catch his horses. As we lost sight of the great millionaire a train of thought arose in the mind of the writer something in this form: "If I had his wealth would I take the chance of being able to catch a horse in a big paddock, where he might be troublesome to catch, and have to hump my own saddle in the broiling sun, or would I have a man at the station to have the horse saddled ready for me when the train arrived?" The answer was, "The man by all means."

An interval of about five years now elapsed before I again travelled with Mr. Tyson. I had been appointed Government auctioneer to deal with the sale of the lands reserved under the Railway Reserves Act of 1877. Some of these lands were part of Jimbour run, and I went to Dalby to hold a sale there.

Sir Joshua Peter Bell was early at the hotel where myself and clerk had stayed overnight, and about 10 o'clock Mr. James Tyson stepped on to the veranda, where Sir Joshua met and shook-hands with him, exclaiming, " Hello, Tyson; where did you sleep last night? You did not camp in the hotel here."

"No; I slept under a gum tree," was the quiet reply.

"Well, you're a queer fish, Tyson, and always will follow the bent of your own ideas," was Sir Joshua's response.

It was quite an interesting study to note the contrast between the two great squatters; both men being of high-class grade in their respective calling-Sir Joshua well educated, well trained in the refined usages of good society, and in all respects an accomplished gentleman; Mr. Tyson, a pure son of the bush, quiet in demeanour, grave countenance, reticent to a refined degree, but possessing a decidedly dignified carriage and manner; a Nature's gentleman. As a matter of fact, I did not know, but it occurred to me as I watched the intercourse between the two men as they stood conversing together on the veranda, that there was a little irritation or disagreement, on some subject. After the sale was over, and when we were seated in the train on our return to Brisbane, the thought of rivalry between the two great men came before my mind. Tyson bought 18,000 acres of Jimbour run; Sir Joshua naturally would not like that, but it is quite possible Tyson did not buy all he would have done had the run belonged to a less popular man than Sir Joshua. Of course, I did not and do not know to the present day, but I have a faint suspicion that there may have been some friendly understanding between the two capitalists not to oppose each, other in their purchases. They were the only purchasers; but as the land brought £1 10s. per acre, I think the State had no reason to complain, for I doubt if the same land has been, or is now, worth the money paid for it then.

The late Mr. Tyson sat next to me on the journey to Brisbane in the train. I cannot explain what there was in the man that attracted; but there was a something that made me feel a deep interest in him and all he said. For an uneducated, self-taught man, he was wonderfully intelligent. As we wended our way along the line, he discoursed upon the geological nature and formation of the country we passed through; and it was very interesting to hear the views of a man who had really learned from the book of Nature the theories and ideas he gave expression to as we travelled on. Both geology and astronomy he had studied in the day and night book of Nature, for the whole of his life was spent both night and day in the "wild" bush. It was quite indifferent to him whether he slept in a house or under a gum tree; as a matter of fact, he preferred sleeping out in the bush to occupying the most luxurious couch in a hotel.

During the trip from Dalby to the city the writer became involved in an argument on the subject of the persistent claim of the squatters to the land, and other incidents in the squatter's life and position in Australia, the outcome of which was the following dialogue, as well as I can remember at this distance of time.

"You know, Mr. -," said Tyson in answer to some remark I had made with reference to his land purchase, "you townsmen make a great mistake in thinking squatters make great profits out of their stock and runs. They are all poor men, and the majority of the runs are really in the hands of the banks, which proves the business don't pay, cheap as you think they have the land."

"You are individually a striking instance of the poor paying character of the squatting industry" I replied with an amused smile.

" Well, but, Mr. -, don't think I have made what little I possess by squatting. I have made what I own in other ways. At one time squatting did pay, but that was when wool brought up to two and three shillings per lb., and the runs had all the best fatting grasses growing in abundance. Now, on most of the runs the blue and other good grasses have become exhausted, and have been replaced by inferior grasses that impregnate the wool with seeds, and thus lowers the price obtainable, and the stock do not fatten so soon, nor do they thrive as well."

"But squatting must still pay," I replied, " or so many would not persist in following the occupation."

"Many do it because they cannot do anything else, and so many are really so deep in the books of the banks that they are practically only managers."

The next station we reached the train stopped for ten minutes. I saw Mr. Tyson rambling about the plain among the grass, every now and again picking up something. When he took his seat, I noticed he had got small bunches of grass between each finger of his left hand.

" Now, look here, Mr. -," he said, addressing me, " here are specimens of some of the grasses. That's blue grass," and then in turn he went on to name the different specimens he had collected, and explained their respective merits; but I have forgotten the names and quality assigned to each grass, but the lecture-for it partook of that character-was both interesting and instructive.

"You think," he continued, "that because I paid you thirty shillings an acre for the land I bought to-day the land is worth what I paid for it. It is not worth more than fifteen shillings an acre, for it won't do for farming; the climate is too dry, and no stock raised upon a capital value of one pound ten per acre can pay. The money would return better interest deposited in the banks at 5 per cent" (the then rate of bank interest for deposits). "I took a fancy to the land, and have paid a fancy price for it."

And so, we chatted away until the train reached Brisbane; the impression formed of the self-made millionaire was decidedly good, for he was a manly man.

The Daily Telegraph, Sydney, Friday, 9 Dec 1898, P 5.

THE DEAD MAN'S GENEROSITY.
A CHARACTERISTIC ANECDOTE.
RETURNING GOOD FOR EVIL

One who knew the late Mr. James Tyson writes, as follows - One evening during my stay at, Felton, Mr. Buchanan, the manager, was called away, and the mail arrived. Mr. Tyson was not good at reading manuscript, and he asked me to read several of the letters he could not make out. Amongst them was one from a gentleman, whom I shall call 'X,' asking for £100; also thanking him for previous obligations.

On my reading the letter, the old gentleman became quite excited, and said: ' I don't know what to do with that man. I will tell you the story. I think it was in the year 1858, I had a big mob of cattle at Bendigo, and owing to some rushes and an increase in the population on this particular market day cattle rose £3 per head. I left all the business to my brother Peter to settle, got a fresh horse and started off up the Riverina, wishing to get there ahead of the news that cattle had risen. At Echuca, I purchased a fresh horse, leaving my tired one. I crossed the Murray, and rode on up the Edwards. I arrived on the Billabong at about 5 in the morning. The horse I was riding was getting very tired, and I found myself weak and hungry. Knowing that there was a station about a mile away, I turned in beside a lagoon, hobbled my horse, and, lit a fire, resolving to call at the station for some rations as soon as they would get up. I knew Mr. X who owned the station by name only. I had never spoken to him, but his wife was a lady who was born in Appin, and her people's place was close to my mother's farm. Her social position was much superior to ours, and we had little or no intercourse; but I remembered that her mother was very kind to my mother at the time of my father's last illness. However, ' Poor fellow me ' (this was a favourite expression of Mr. Tyson's) went up to the house, and knocked at the door, and Mr. X came out. I bade him' Good morning,' and told him that I was travelling and asked him if he would sell me some rations. He was a very proud man, with a military training. He turned on his heel, and. said, 'No, we don't keep an accommodation house for you travellers; you should provide for yourselves.' I said to him ' Look here, Mister, I am very far gone, and have 30 miles to go before I can get anything; if you do not give me something, I will not be able to go on.' He knew that I owned Tupra, and had often seen me driving cattle. He kind of repented, and said, 'Go to the back, and the cook will give you some broken bread if he has any.'

The old saying that ' Hunger has no pride ' was true in my case, and I went to the back. Mr. X sang out 'To give me some scraps.' The cook was a good sort, and he wrapped up a loaf of bread and a shoulder of mutton, with a pinch of tea and sugar. I astonished him by giving him half a sovereign. I went down to the river, boiled my quart pot, had a feed of the bread and mutton, went on, bought a thousand bullocks, and did well out of them. Mr. X was a proud extravagant man and not a good manager. He lost his Station, and entered the Government service in a neighbouring colony. I never heard any more of him until last March, when I received this letter.

' Going to a drawer and taking out a letter, Mr. Tyson handed it to me to read, and it ran thus: — 'Dear Mr. Tyson, — Necessity compels me to make this application to you, which I hope you will not refuse. I certainly have no claim on you but I ask you to remember the old pioneering days in the Riverina, and, further, that my wife was born at Appin like yourself. The fact is that I was retrenched some years ago, and have been unable to find employment. We are very badly off, and if I cannot raise £100 by the 4th of next month, we will lose everything we have in the world. £100 is not much to you, but it is all the world to us. — Yours faithfully, X.

Mr Tyson continued: 'I wrote out a cheque for £100, and sent it to him without any letter. A few months after another application came from the same gentleman for another £100 appealing on account of his wife's illness. Remembering that she was reared a lady, and that her people were very good to my mother, I sent the second £100, just enclosing the cheque in an envelope, as before. Now he has applied a third time. What shall I do?'

Wishing to see what, direction the old man's action would take, I said to him: ' You are the best judge of that, Mr Tyson.'

He considered for some time, and said, 'I will send him £50. Will you write me a letter saying he must not ask me anymore?'

I wrote him a note, stating that enclosed was a cheque for £50, and under no circumstances was the applicant to apply again, as it would cause the pain of a refusal.

The old gentleman wrote a cheque,' and when he was about to fill in the figures he stopped and said, ' I think it had better be £100; £50 might be no good to them, and they are a very old couple now.'

CHAPTER 5

Statements by witnesses in the trial to determine James' domicile

After James Tyson's death the Queensland Government had a trial to determine where James' home was because, if it was Queensland, some of his assets in the other colonies would be subject to Queensland death duty.

In the trial, as well as matters directly related to his domicile, some witnesses gave additional information about James. In board terms the official court records only included matters which helped in determine James' domicile. This would contribute little in my quest to understand James. To me it is of little consequence in which of the many houses he owned he lived.

Two newspapers, The Brisbane Courier and The Telegraph (Brisbane), had a reporter at every day of the trial and reported the proceedings the following day or, when they produced an afternoon edition, the same day. These reports had valuable information about James and it is all under sworn testimony by people chosen because they knew him. This removes one of the problems with articles written by authors who had never met James but gave the impression that they had.

There were 94 witnesses and of those only half, 47, were recorded in either of the newspapers as detailing items of interest other than regarding domicile and that is the number I have included in this publication. When both newspapers had information from a witness substantially the same, I have only included the report which was the most explicit. There were several times when the newspapers reported different parts of a testimony and I have included both. There were enough instances of each paper reporting the same piece of testimony for it to be shown that neither was fabricating information as I believe both are right. There were other reasons for the difference; one paper may have felt some of the testimony may not have been newsworthy or it could be simply due to a shortage of space.

It has been suggested that the reporters' own knowledge of James Tyson had some influence on their report. One of these items that is now in folk-lore is James never used soap but he used sand instead. I did an exhaustive search before the trial of any mention of James not using soap and I found nothing. Therefore, I believe that anecdote arose from the testimony at this trial. It would be not unexpected that a person researching now could deduce that it was well-known in the community before the trial without a deliberate search before and after the trial.

Comments by:

Witness: Robert Bourne, Manager & Inspector of Telegraphs, Sydney

***Brisbane Courier,* 16 Nov 1900, P 7.**

Witness met deceased on the Tully selection and they had a conversation. A remark was made about friendship and deceased told him that he never had time to make friends.

Witness: William Boyd, Editor, Queensland Agricultural Journal
Telegraph, **20 November 1900, P 2.**

Deceased said at the union Club in 1883, he kept a journal and showed him a pocket book, which, when filled, would be left with others at Felton, witness said he would like to get them to write his biography but deceased said he would like to write it himself.

Witness: Henry Bracker, grazier and manager of the stock department of Moreheads Limited,
Telegraph, **28 November 1900, P 1.**

Witness stated his firm were agents for deceased. Deceased had frequently remarked to witness, "Home is where night finds me." This was also a common expression of his, "I am a wanderer and a gatherer of riches." Witness in referring to the long journeys undertaken by the deceased had said to him, "If you do not take more care of yourself, Tyson, you will be found dead on the bank of a gully like an old working bullock." Deceased did not reply, but witness noticed that he changed colour and went pale. He had spoken of selling Felton and a few years before his death he had said, "I would like to sell all my properties and take a rest."

Witness: James Brown, Cattle Overseer, Meteor Downs
Telegraph, **30 November 1900, P 2.**

Witness entered deceased's service in 1888 and 2 years later he presented him with a gold watch and chain for his sobriety and general good conduct. Witness produced the watch with that inscription. Deceased told him that he was always travelling and could not settle down. Nobody else could do the work he does.

Brisbane Courier, 30 November 1900, P 7.

Witness entered Tyson's service in 1888 and two years later he was presented with a gold watch and chain "for his sobriety and general good conduct as a drover." Witness produced the watch.

His Honour: Is it a keyless watch?

Mr Lilley: Yes, your Honour

His Honour: Tyson used to say that every sensible man should use a keyless watch. I remember him telling me that he doubted if I was a sensible man. I asked him why and he said, "because every time you open your watch to wind it some dust gets in."

Witness (continuing) deposed that Tyson spoke of himself as a man who was always travelling and lived in no fixed place.

Witness: James G Buchanan, Manager, Felton
Telegraph, 28 November 1900 P 2.

Mr. Buchanan, continuing his evidence, stated that deceased told him that when he (deceased) secured a good property and improved it he always left it in the hands of his managers. Deceased did not normally stay at Felton after he had finished his correspondence. The longest witness had known him to stay there after the correspondence was completed was seven or eight weeks. He had told witness that he was always glad, to get away because his correspondence bothered him. He told witness that on several occasions he had spoken about the place of his birth. About 18 months before his death, he had told witness that he looked on the place where he was born as his home. At Christmas time in 1896 deceased asked if he (witness) was going home. Witness replied, " I do not know?" Deceased said that until recent years he had made a habit of going home himself. He used to go to Wilton to visit his sisters. Deceased assisted his sisters by sending them money. He never told witness that he regarded Felton as his home. He had said, " I am a bird of passage. I have no home." In 1897 deceased spoke to witness about disposing of Felton. Witness had arranged for the men to commence work at a dam when deceased said that he intended to let the Government have Felton, and there was no use doing any more work on the place. Some years back he had heard deceased say that he would let the Government have Felton if they would give him five acres to one of western lands. Witness had tea with deceased the night before he died. He did not then appear to be very ill. Witness thought his death was due to pneumonia. No medical man saw the body after the death of deceased. On two or three occasions deceased told witness that the climate of the Downs was too cold for him in winter time. The cost of Felton house was about £1,200. A house was erected on Mount Russell at about the same cost. Mr. J. Tyson Doneley brought some pear trees to Felton in the year 1878. Witness understood they came from deceased's mother's garden in New South Wales. Witness remembered deceased giving evidence at the criminal sittings of the Supreme Court at Toowoomba when a person was charged with stealing sheep belonging to deceased. In his evidence deceased stated that "he resided at Felton at present and at other places at other times."

Cross-examined by Mr. Woolcock: Witness laid the information in the case and described the sheep as belonging to Mr. James Tyson, of Felton. Another case was tried in Toowoomba in 1890. Witness did not remember what deceased said about his residence on that occasion. Witness then described the sheep as belonging to Mr. James Tyson of Felton. Two men were charged with stealing deceased's cattle in 1895, and the case was heard at Ipswich. If deceased swore that he was a grazier, residing in Queensland at that time it would have been true. Witness could not remember any year while with deceased when it would not have been true. Witness had signed a declaration that deceased had continuously resided on freehold land at Felton, for a period of two years, from 1871 to 1873. Felton was deceased's home while he resided in Queensland, which was up to the time of his death. Witness never heard deceased use the word "home," in connection with Felton. He had used it in letters. Deceased wrote his letters on slates, and witness used to copy the letters, and read them to him. Deceased would not allow witness to keep his wife at Felton, as he objected to women being about the place. Felton was the only station belonging to the deceased where the manager was not allowed to have his wife residing with him. Witness thought the library at Felton contained about 100 volumes. Witness told the Crown solicitor that deceased if at all sick made for Felton, it was true. The impression on witness's mind was that in his later years he came to Felton to rest.

Mr. M'Donald had forwarded letters to Felton from Tinnenburra. Deceased received many begging letters. In some he was asked for £100 or £1,000 straight out. Witness destroyed a lot of letters.

Mr. Woolcock: Without consulting deceased?

Witness: Yes.

Mr. Woolcock: How did you know whether they were of any value or not?

Witness: I used to destroy them. Some were offers of marriage.

His Honour: Did you not show him those?

Witness: No. He would have been annoyed.

His Honour: Who at?

Witness: At me for showing them to him.

Continuing, witness stated that deceased had not spoken to him of building a grand house at Felton. He spoke of building a " humpy," which with him would be a substantial cottage. Deceased regarded Mount Russell as the best of his properties. Deceased might have told witness that he would not put in another winter on the Downs, and that he would go to Meteor Downs. Witness did not see Mr. Tyson Doneley bring the pear trees to Felton. He had heard deceased say in 1896 to Mr. Tyson Doneley, "You brought those trees, Doneley?" Mr. Tyson Doneley replied, "Yes." Witness did not know anything about deceased bringing trees to Felton. Of late years deceased had spent his Christmas at Felton. He had spent one Christmas at Mount Russell, and three 'or four times he had gone to New South Wales. Witness generally stayed at Felton on Christmas Day, but it was not to keep deceased company. To go home witness had to ride 50 miles, that was into Toowoomba, and it was too much to do in one day. There was only one holiday (Christmas Day) at Felton. Sunday was devoted to letter writing.

Mr. Buchanan, further cross-examined by Mr. Woolcock, stated that he did not know who paid Miss Hewitt's wages while she was housekeeping at Felton. Witness looked upon her as deceased a housekeeper. Witness received £200 per year salary when he went to Felton, then £150, and then £120.

Mr; Woolcock: Your salary did not increase with long service?

Witness: No, it decreased.

Mr. Woolcock: You were getting £120 per year as manager at the time of Mr. Tyson's death?

Witness: Yes.

Mr. Woolcock: And you had to keep your wife and family in Toowoomba?

Witness; Yes.'

Mr. Woolcock: How much are you receiving now?

Witness: £200.

Mr. Woolcock: When did you get the increase?

Witness: In June, after the death of the deceased.

Mr. Woolcock: It was given you by the administrators? You are now receiving £200 and you have the privilege of keeping your family at Felton?

. Witness: Yes.

Mr. Woolcock: Were you, offered anything for giving evidence in this case?

. Witness: No.

By Mr. Lilley: Why was your salary reduced?

Witness: To cut down expenses.

Witness: Robert Bushe, Solicitor, Sale

Brisbane Courier, **24 November 1900, P 12.**

Witness said Tyson used to speak of himself as a wanderer.

Witness: Mr. Charles Hardie Buzacott, M.L.C., former editor of the Brisbane " Courier"

Brisbane Courier, **Friday 16 November 1900, P 7.**

Witness said he knew Tyson fairly well. In December, 1893, Tyson gave him an interview which was published in the " Courier" of the 8th December in that year. Five or six weeks afterwards Tyson saw him, shook him cordially by the hand, and said, " It's all right, mister; I believe it's done good; I believe it's done good." He had not met Tyson in the meanwhile. Tyson had granted the interview reluctantly, and appeared up to that time to avoid witness.

Mr. Feez tended a newspaper cutting containing the interview in question.

Mr. Lilley objected that it was irrelevant.

Mr. Feez directed his Honour's attention to the references in it to climate, to Tyson's residence as being " Felton, Darling Downs," and to Tyson's views on the border tax.

His Honour admitted the printed interview.

The examination of Mr. Buzacott was resumed. Witness had described Tyson's place of residence as "Felton, Darling Downs." He did not remember if Tyson told him so, but he personally had always understood that Tyson's home was at Felton. At any rate Tyson found no fault with anything in the interview.

By Mr. Lilley: In the interview Tyson did not refer to Felton particularly. The general trend of the interview was that despite the drought the prospects of Queensland were good. Referring to the Western country, he said we could rear the best sheep, the best cattle, and the best men in the world there. Tyson gave witness the impression of being a man who belonged to the whole of Australia.

Witness: William Castles, Auctioneer, Brisbane
Brisbane Courier, **20 November 1900, P 7.**

Tyson had visited him, and he had visited Tyson. In 1884 he stayed five days at Felton, which Tyson said was his home. The deceased spoke of the station as most conveniently situated to his other properties, north, south, and west. He said he would exchange it, however, for land out West, if the Government agreed to his proposal of four acres to one. Tyson had already picked upon the land—out Longreach way—and talked of making it his home if Felton were exchanged. Tyson told him that when away from home he travelled under the name of Smith. Also told witness that he never married, and mentioned the reason, and that he had never made a will, and never would. His property, he said, he would handle himself during life. He expressed his intention of doing something for the youth of the colony, for our future legislators, and of contributing his share towards, "maintaining an army for the defence of our homes, because," be observed, " this colony is going to be great, rich, and valuable."

By Mr. Lilley: Tyson had told witness many things; among others, how he laid the foundations of his fortune at Bendigo, where his brothers sold the meat, he sent down to them from the western districts of New South Wales, and where they stored their money and notes in a washtub, and only took it to the bank when full to the brim.

His Honour: Only that I know Tyson was a teetotaller, I should say he had been drinking when he said that.

Mr. Lilley (to witness): Don't you think Tyson was a bit of a romancer? -that he was fond of pulling people's legs?

Witness could not say that he was.

If Tyson told you that he got £300 a head for horses, would you believe him?

If he said that I should be inclined to think: he was a bit of a liar. (Laughter.)

Telegraph, **20 November 1900, P 2.**
Witness resided at Loganholme. Met deceased in 1881, when deceased asked him to make a valuation of some farms at Loganholme. Witness stayed with him and had many conversations with him afterwards. Deceased said Queensland was the best of the colonies. In 1884 witness visited Felton and at that time asked deceased where his home was. Deceased said, "Felton is my home; it is most convenient for me, I am near, the railway at Cambooya and from there I can go to any place." Deceased was fond of Felton, but said he would be willing to exchange it with the Government if they would give him four acres to one of certain land he had chosen. He added, "There I will make a home to finish my days." Witness asked him if he did not get the exchange did he intend to live and die at Felton, and he replied in the affirmative. Felton was a comfortable bachelor's home, and there was a fair library. Deceased told witness

to write to him at Felton. He also told witness that if he was away from home to look among the passenger list of the steamers going north for " Mr. Smith," and that would be him (Mr. Tyson). Deceased gave witness his reason for not marrying, and said he would never make a will. He said, "I intend handling my property during my life time, and I will do something for the youth of the colony, from whom our legislators are to be made." He further added, " I will do my best to get up an army to defend our colony." (Pointing out that the colony was likely to be very rich and prosperous). Deceased arranged with witness's son to go to Felton in 1894, or as he termed it to go home with him."

In cross-examination by Mr. Lilley witness stated that he did not think deceased was a bit of a romancer.

Mr, Lilley; You took everything as mother's milk?

Witness: No.

Mr. Lilley; Do you not think that among all his. virtues he had a habit of telling lies?

Witness: No.

Mr. Lilley; Do you not think he had a habit of pulling people's legs?

Witness: I do not think so.

Witness: Mr. Henry St. George Caulfield, Polynesian inspector
Brisbane Courier, **Tuesday, 20 November 1900, P 7.**

Witness deposed that Tyson told him on one occasion, when witness was his guest at Felton, how he managed his numerous properties, dropping in upon his managers by surprise, and always returning to Felton. When witness visited Felton, Tyson greeted him with those words at their first meal " You are welcome to Felton; make yourself at home and there is no starch."

By Mr. Lilley: He gathered from Tyson's conversation that he was a wanderer, but not that he had no fixed place of abode. Tyson told him distinctly that his nephews and nieces were to inherit his money, and that if they derived as much pleasure in spending it as he had obtained in making it, he would be satisfied. Witness told him that he might rest satisfied. (Laughter).

Telegraph, **19 November 1900, P 5.**

Witness visited Felton in 1883 to see deceased on business matters. Deceased in course of conversation stated that he supervised all his properties from Felton. At the first meal witness had there Tyson said, " You are welcome to Felton - make yourself at home; there is no style."

We have in the above testimony the only item of contention between newspapers covering the trial. Did James say, "...there is no starch" or did he say, "...there is no style"? It could have been either because James is reported to have deleted starch from a list of expenses because he classed it as a "luxury" and, therefore unnecessary, which, may have been discussed at the table during dinner mentioned above. James would not be in favour of anything "stylish" as it is close to what he calls "lah-de-dah" women of which he is on record saying he abhors. One of the reporters mis-heard what Mr Caulfield said and we will now never know what James did say. But who cares? The intent is similar with either word; that there was no pretence at Felton when James was in charge.

Witness: Edward Danaher, Messenger, Crown Solicitors
Telegraph, **22 November 1900, P 2.**

Witness stated that he had counted the documents found at Felton by Mr. J B Hall, curator in intestacy. These were chiefly letters and telegrams and were mostly addressed Felton. The total number he counted was 9,080.

Witness: Henry Daniels, ex-M.L.A., Cambooya
Brisbane Courier, **13 November 1900, P 2.**

Witness first met Tyson twenty-one years ago on the Warrego. Tyson was then on his way back to Felton and said he was in a hurry to get home. Witness had frequent conversations with him afterwards. On one occasion, talking about Felton, Tyson remarked that, when he first he saw the place, he determined to buy it and make a home of it for his old age. He told witness how he came into possession of the station. The previous owner had lived beyond his income, borrowed money from Tyson, and then " went travelling on the Continent with Lord this, and Lord the other, and Lord knows who." (Laughter). As a consequence, he could not pay the interest on the loan, and Tyson gave him £30,000 or £40,000 more to clear out altogether. He believed the total cost was over £100,000. Tyson had told him that he liked Felton because the climate agreed with him, save in the winter, when he liked to go north. He always referred to Felton as his home. Witness met him in Brisbane about two weeks before his death. He was looking ill, and said so. Witness asked if he was going to stop, and he replied, " No; I am going home to-morrow." He wanted witness, to go with him, and stay a week in pursuance of long-standing promise, but witness feared his motives might be misconstrued, and declined. About two years earlier Tyson said to him, "'I am very lonely. I have never injured a man knowingly in my life, and I am over 70 years of age, and haven't a friend in the world. I have any number of relations, and I believe they want to know when I'll die, so that they can get the money. Witness re-joined, "You may be wronging them. I don't believe they are all that way, for I know some of them myself." Tyson answered, " Well, it may be so, but I feel the only real friend I have is here" (patting his trouser pocket). " All the same. Daniels,'" he added, " when you get old you feel you would like someone to pay you some slight regard. I get very lonely at times. I always have to get a lot of these" (abstracting from his coat pocket a bundle of " Ally Sloppers," " Tit-bits," " Scraps," &c.) "to divert my thoughts." Witness once asked him how he managed to keep a grasp of the whole on his affairs. " It's the simplest thing in the world," he replied. "I have never showed it to anybody, but I'll show it to you." He then drew four pyramids on a leaf of his pocket book.

These represented his properties in New South Wales, Victoria, and Queensland. "The four of the largest represented," he said, " Mr. Stokes; that is, myself." He went on to explain. " I make my bank managers do my bookkeeping, and I don't have to pay for it. When I send sheep or cattle from one station to another, the station to which they are sent must pay a cheque to the other station for the current value of the sheep or cattle, and of course when the cheque goes through the bank the bank has to keep account. Sometimes, when owing to improvements, droughts, &c, a station doesn't pay expenses, it has to borrow from Mr. Stokes. On the other hand, after the working expenses of each of these stations have been paid, all the surplus revenue is placed to the credit of Mr. Stokes. The banks have to keep account of the transactions, and see that the amounts correspond with the accounts of the various stations." Witness told how on one occasion Tyson, to save exchange, drew £80,000 in sovereigns from a New South Wales bank and brought it to Queensland with him by boat. This anecdote reminded his Honour that a jury once found on oath that a lady had carried 20,000 sovereigns in a belt round her waist for a lengthy period. Continuing, witness said that Tyson used to draw £25 a month for personal expenses. Once he complained that he had been extravagant, inasmuch as he had drawn £20 over his allowance.

Attorney-General: Did he take much interest in local politics?

No, very little, he was too busy making money.

Attorney-General: Did he attend to his Legislative duties regularly?

No, he did not care about them. He had intended to attend Parliament regularly. But Mr. Macdonald-Paterson induced him on one occasion to make a speech on the marsupials of Queensland, and after that his interest in the House fell off. The reason was that in the middle of his speech when he was describing how wallabys (sic) came for thousands of miles out of the scrub into the paddocks and ate the grass, Mr. Perkins interjected " Do they bite the sheep?" Tyson was so disgusted that he sat down, and would never speak again. Moreover, he complained that he could not hear well, and when he did hear he could not understand, owing to the acoustic defects of the Chamber.

Mr. Lilley: Do you think Tyson was a man who would give £200 to a station owner for his equity of redemption, and then tell you he gave £30,000 or £40,000 for it?

Witness would not say that, though he knew Tyson was not exactly Truth itself. On the question of celibacy, witness considered Tyson was a bachelor for the reason that on one occasion towards the close of his life he complained of being lonely, and said he would perhaps have been happier if he had married in his young days and had his family round him, " he would not have been so rich, but he might have been happier."

Telegraph, 13 November 1900, P 3.

Witness stated that he met Mr. Tyson 21 years ago on the Warrego. The deceased was then returning to Felton, and said he was in a hurry to get home. Witness had frequent, conversations with deceased afterwards. On one occasion, when talking about Felton, deceased remarked that, when he first saw the place, he determined to buy it and make a home there. Deceased told witness how he became possessed of the property, and the amount it had cost him. The previous owner, he stated, had borrowed money from him and was unable to pay the interest, consequently he (deceased) had given him £30,000 or £40,000 more and taken possession of the property. Deceased' said the climate of Felton agreed with him except in winter, ' when he liked to go north. Deceased always referred to Felton as his home. Witness met him in Brisbane about two weeks before his death. He was looking ill, and told witness he was not well. Witness asked him if he was going to stop, and he replied, " No; I am going home to-morrow." He asked witness to go with him and stay a week, but witness declined. Deceased said, ' "I am very lonely. I have never injured a man knowingly in my life, and I am over 70 years of age, and haven't a friend in the world. I have any number of relations, and I believe they want to know when I'll die so that they can get the money." Witness replied, "You may be wronging them. I don't believe they are all that way for I know some of them myself." Deceased answered, "That may be so, but I feel the only real friend I have is here," patting his trousers pocket. " All the same Daniels," he added, "when you get old you feel you would like someone to pay you some regard. I get very lonely at times. I always have to get a lot of these (producing some comic papers) to divert my thoughts." Witness asked deceased how he used to keep grasp of his money matters, and deceased replied: "It is the simplest thing in the world." Witness explained a plan which he had for dealing with his properties in each of the colonies, and how he dealt with the banks. He told witness that he used to draw a cheque of £20 every month for expenses, and one week he said he had been extra extravagant, when he had drawn £20 more. Witness remembered deceased meeting with an accident at Leyburn in May, 1898. Witness then saw him at Felton, when his face was very much cut. Witness asked him how he met with the accident, and deceased replied that about a week previous he had started out from Felton and at Leyburn had run over a stump. Next morning, he stated he was so sore from the accident that he had to turn round and come home. Witness remembered deceased's niece, Miss Hewitt, keeping house for him. Deceased did not take much interest in the affairs of the Felton district—he was too busy making money. He was a member of the Legislative Council, but he did not attend very regularly. He had made up his mind to be regular in his attendance and was, until he was persuaded to make a speech on a marsupial bill. He was then describing how the scrub wallabies came out or the scrub, when the Hon. P. Perkins asked him, " Would they bite the: sheep?" "What?" said the deceased, and he afterwards left in disgust at thinking that he had got into a place where they were so ignorant that they did not know whether a wallaby would bite a sheep or not. (Laughter.)

Cross-examined by Mr. Lilley: Deceased often spoke of his properties in New South Wales and Victoria. He did not tell witness about his mother and brother dying in New South Wales, or that there was a family vault there. He did not tell witness that he was a wanderer and always on the wing. He never spoke to witness about selling Felton to the Government. He had stated that if the Government promised sufficient inducement he might "swop " all his Darling Downs property and take in exchange some, western lands.

By the Attorney- General: Do you know whether deceased was single or married?

Witness: Single

His Honour: Did you ever ask him if he was married?

Witness: No; but he led me to believe he was not. In the conversations with witness, he used to say, "It would have been better had I got married in my young days. I would have had a grown-up family now, and someone to take care of me. I would not, have been so rich, but I would have been happy."

By Mr. Lilley: All the conversations witness had with deceased were in Brisbane, going to Felton, or at Felton.

Witness: JAMES H Davidson, Grazier Westbrook
Telegraph, 29 November 1900, P 2.

Stated that he knew deceased and had frequently conversed with him. Deceased had told witness that he was always wandering and that he inspected all his properties himself. He had said that he would exchange Felton for land out west, and he and witness tried to work out a basis of exchange to submit to the Government. On another occasion deceased told witness that he had been to Sydney to try and purchase a property, but that the price was too high and the taxation too heavy.

Cross-examined by Mr. Feez: Had received letters from deceased, and had written to him. The letters were addressed from and sent to Felton. As a resident of the Darling Downs, witness knew that deceased's headquarters were at Felton. Deceased took an active part in the affairs of the Jondaryan Divisional Board, and was the moving spirit at its formation. Witness would consider that deceased was a resident of the Darling Downs.

By Mr. Lilley: After the formation of the Jondaryan Divisional Board deceased did not take much interest in it.

Witness: Hon J F G Foxton, Home Secretary, Queensland
Telegraph, 17 November 1900, P 2.

Witness stated that he met deceased on an occasion about five years ago at the Queensland Club. Had known him from the year 1878. In conversation with deceased at the Queensland Club, he stated that someone had offered him £5 for his watch guard, which appeared to witness to be a piece of boot lace. He also spoke of a gentleman having offered him £100 if allowed to ascertain his system of bookkeeping. Deceased said, " The joke is I have no system. I make the bankers keep my books." Witness had been under the impression that deceased had an office in Melbourne, and asked him if such was the case. Deceased answered in the negative. Witness further asked, "Where are your headquarters?" Deceased replied, " My headquarters are on the Darling Downs. Felton is my home." Witness met deceased afterwards in the train, and he then looked very ill. Witness advised him to see a doctor, and deceased replied, "I will soon be all right now as I have got home, and will have better attention."

Witness: Chas. Freestone, Constable, Brisbane
Telegraph, 1 December 190,0 P 2.

Witness stated that he had lived at Cambooya until 1897. Was 22 years of age. On four occasions had started from Cambooya to take deceased to Felton. Twice witness had ridden all the way with him and twice they had been met by a boy. The last occasion was at the end of 1896. Deceased told witness then that he had been on a long trip round his stations, and that he was ill and "done up." He further told witness that he did not think that he (witness) would accompany him again from Cambooya, as he was getting old and he was afraid to visit his stations in case he should not come back to Felton, his home. He also said that Felton was always, his home, and that was where he wished to die. Deceased did not look very well at the time.

Witness: John Fuller, Farmer, Southbrook
Telegraph, 13 Nov 1900, P 6.

Witness stated that, he had been about 40 years in the district. Knew the late Hon. James Tyson well. Witness was boundary riding on Eton Vale station, which adjoins Felton, at the time when deceased came to Felton. Witness had worked on Felton for deceased. Had many conversations with the deceased. On one occasion in 1885, witness was working at the erection of a new wool-shed at Felton, when deceased referred to the journeys he undertook, and the distances over which he used horses. Deceased had a great opinion of Felton. He generally used the expression, "I will be home on Saturday."

Cross-examined by Mr. Lilley: Deceased told him in 1893 that he would sell Felton if he got a satisfactory price. Deceased then appeared to take an interest in the village settlements, and said if the Government wanted it for that purpose, he would sell it. Deceased did not take much interest in the affairs of the Felton district.

His Honour: He took an interest in local taxation.

Mr. Lilley: Yes; he took an interest in everything that touched his pocket.

Witness: Hon. A C Gregory MLC & Surveyor
Telegraph, 29 November 1900, P 7.

Witness stated that he had been Surveyor-General of the colony. He had frequent conversations with the deceased. The latter had alluded to Felton in a general way when speaking of his properties. In 1876 witness had said to him, "Have you got a main camp?" Deceased said, "No, I never had but one home." He said that was on one of the New South Wales rivers — witness had forgotten the name. He further said that he had made up his mind to build a house there, and that he had cut a deep channel from the river to bring the water through a " billabong." Witness knew that deceased resided at the house afterwards. On another occasion witness said, "I suppose, Tyson, you are something like myself. You are a wanderer on the face of the earth. We are like an emu, which has no fixed habitation, and which is only to be found where the last showers of rain has caused the grass to grow." The reply of deceased was in full concurrence with the remarks.

Witness: Henry L Groom, Bus Manager, Toowoomba Chronicle
Telegraph, 16 November 1900, P 2.

Stated that he saw deceased occasionally in Toowoomba. Deceased generally came into town from Felton. Remembered deceased coming in about the Chronicle some years ago. The paper had been addressed to " The Manager, Felton," and he desired that it should be addressed personally to himself, " Mr. James Tyson, Felton." Deceased often came in to see Mr. W. H. Groom, M.L.A., and if the latter was engaged, while waiting he would enter into conversation with those in the office. Witness often heard deceased say, " I am back home again." Witness had approached him regarding a subscription for a rifle association. Deceased said he would be pleased to encourage rifle shooting, and gave a donation of 20 guineas. He further stated that he would further contribute if necessary, and told witness to address any correspondence to Felton. He added, "As I am living here, I would like to assist everything local." Deceased took a big interest in the Defence Force. He often gave prizes anonymously to be competed for in rifle shooting. He gave another prize of £100 for a competition open to young ladies of the district.

Cross-examined by Mr. Lilley: Sometimes addressed correspondence to the deceased in Brisbane when an immediate reply was wanted.

Witness: W H Groom M.L.A.
Telegraph, 19 November 1900, P 2.

Witness stated that when he went to Felton, he saw deceased, who invited him to remain there. He said, " I will not stay, as the people will say I talked you over." Deceased told witness some time afterwards that it was not the value of the land he cared about, but he wanted to consolidate Felton, and make it his home and headquarters. "Witness travelled in the train with the deceased at the time of the departure of the New South Wales contingent for the Soudan. Deceased showed witness a cheque for £2,000 which witness believed was marked "

Felton," and said, "That is my contribution to the New South Wales patriotic fund." Deceased had given several contributions to witness to help local objects in Toowoomba. Witness's experience of deceased was that he was very liberal, in fact exceedingly liberal in many ways. He never refused any request witness made. Deceased regarded Queensland as the best of the colonies, and that Meteor Downs and Felton were the pick of his stations. Deceased had once interviewed Sir Thomas M'Ilwraith with regard to buying 1,000,000 acres of Tinnenburra run at 5s. per acre. Sir Thomas M'Ilwraith said the proposal would have to be sanctioned by Parliament and referred deceased to witness. Witness told him that he could not think of bringing the matter forward, as it was entirely, against his 'principles! On one occasion, in 1895, witness had a conversation with the deceased in the train, when witness said his son would like to get a situation on Meteor Downs station Deceased said, "Well, I am going home to Felton; when I get to Cambooya I will send a wire to Mr. Brown, and if there is a vacancy, he will take your son." Witness afterwards received a telegram from deceased, as follows: " Off to Sydney, just received wire from Brown, who will take your boy." Deceased used to tell witness his business in confidence. Witness knew that deceased was going to Sydney at that time to see about buying Lassetter's property with the money which he was about to receive on matured Treasury bills in Queensland amounting to £200,000. Deceased told witness that after giving his contribution to the patriotic fund in Sydney he had received 60 letters from persons claiming to be related to him. One of these pointed out that deceased was related to the Duke of Northumberland and asked for his portrait lo help to establish the claim. These letters were received by deceased at Felton, and he had stated that he had only answered one of them.

Cross-examined by Mr. Lilley: Was certain that deceased used the word "home" when speaking of improving Felton. He used the word "home" frequently in his conversations with witness. Deceased did not care much for politics. Deceased never spoke of exchanging Felton for a large area of freehold in the western districts. He told witness that

Heyfield, in Victoria, was one of the best fattening properties he had. He further said that before he saw the Darling Downs he had chiefly resided at Riverina, in New South Wales.

Witness: James Boyce Hall, Curator of Intestates Estates
Telegraph, 17 November 1900, P 2,

Witness stated that in December, 1898, he was appointed administrator of the deceased's estate under the Intestacy Act. As administrator he paid a visit to Felton. Witness made a thorough search of the premises, more particularly in looking for a will. Felton was substantially built, well appointed, the rooms were of fair size, and he would describe it as a comfortable home for a bachelor. There was a private office, which was locked, and there was also a very fair library. Witness described how the papers were found in the office. Those were strewn about on the shelves, or lying on the floor. Amongst the papers there were many documents and letters which had been tied up in bundles and thrown into the corners of the room. Witness found a number of diaries in the private office. These dated from 1854 up to the year of deceased's death. The letters witness found were in their original envelopes.

His Honour: Stamps should be a valuable asset in the estate. Some of them should be worth £5 apiece.

Mr; Feez: The envelopes and stamps are all gone now.

His Honour: Well, they should have been worth a good deal.

Witness, continuing, said he found letter books, account sales from different auctioneers and station managers in the manager's office. Amongst some valueless papers was found a fixed deposit of £30,000 in the Queensland. National Bank, scrip for 977 shares in the Queensland National Bank, and a fixed deposit receipt of £60,000 in the City Bank; also, a deed of grant of land, and a paper appended to it about a loan of £100. Witness put all the documents into seven tin boxes, and brought the more valuable of them to Brisbane. When the papers were investigated at Felton they were in their envelopes, but were not put back again. The envelopes were not destroyed. Most of the letters were addressed to " Felton, Cambooya." Witness, afterwards handed the papers to the Queensland Trustees Limited, when the latter were appointed administrators. The envelopes were placed in the boxes with the other documents. some southern newspapers and a lot of local newspapers were found. Many were addressed to deceased at "Felton, Cambooya." There were also letters from New South Wales and Victoria, which were addressed to Felton.

Cross-examined by Mr. Lilley: The bank deposit receipts had expired. The papers were kept in fairly good order in the manager's office. Apart from the diaries very few papers of any value were found in the private office.

Witness: F X Heeney, Under Secretary for Lands
Brisbane Courier, **17 November 1900, P 14.**

Witness said he made Tyson's acquaintance when acting as land agent at Toowoomba in 1870-73. Deceased told him he regarded Felton as the best of his properties for residential purposes, the climate suiting him better than elsewhere, and that he intended building a residence there. In 1875 Tyson spoke of contesting the Warrego electorate for a seat in the Legislative Assembly.

By Mr. Lilley: Witness did not recollect Tyson ever saying that he had no home in the ordinary sense of the word, that he was a wanderer.

Did you ever hear him say, " I am off on the wallaby track again, following the same old star?"

Witness could not remember if he said that, but he had heard the deceased speak of a star—a star which he followed. Tyson was a peculiar man. When he started on a journey, he said he never turned to look behind.

The Attorney General: Dr. Johnson never passed a post without touching it. It was only a fad.

Witness further cross-examined, said Tyson had spoken to him of Heyfield, in Victoria, but said it was too grand, that there was too much style about it for him.

Witness: Isaac Henry, grazier, Bellenden Plains NSW
Brisbane Courier, 30 November 1900, P 7.

Witness deposed to conversations he had with Tyson at various times and places the latter said, he had made preparations for his burial near the place where he was born that, though he travelled thousands of miles, the place where he was born was always home to him; that wherever he went he was at home; that he had a home at each of his stations, and that Cobb arid Co's coach was home to him during a great part of his time.

Witness: Mrs Annie C Hogarth, Widow, Balgownie
Telegraph, 15 November 1900, P 7.

Witness had invited him to dinner one Christmas Day, and she had received a reply written by his manager declining, and alleging as a reason that deceased considered every man should spend Christmas Day in his own home.

Cross-examined by Mr. Lilley: She was not sure whether the word home " or " house" was used in the letter sent declining her invitation. She had asked deceased why he did not take a trip to England, and he had replied to the effect that he could not leave his business matters. She suggested that he should go at the time of the diamond jubilee, but the objection he raised was that the time was not suitable and there would be too much of a crowd. Deceased failed very much during the latter two years of his life.

Witness: Frank Alex Huet, Dentist, Rockhampton
Telegraph, 9 November 1900, P 2.

Witness stated that he was acquainted with deceased. First met him in Toowoomba about the year 1892. Witness had about three quarters of an hour conversation with him on that occasion. Witness met him frequently afterwards when travelling between Brisbane and Keppel Bay. Deceased used to seek witness's company; witness had formed the opinion that deceased was an eccentric old gentleman and used to crack jokes with him. On one occasion, witness met deceased at Emerald, on the Central line. Witness believed the date was 1896. They were together the whole of the Sunday. Witness discussed the question of his home with deceased. Witness said, "Well, you tell me that you travel many thousand miles during the year, and I believe you are a bachelor. I presume you have no fixed home?" Deceased said, "Why do you ask that?" Witness replied that he considered it would be an advantage to a bachelor not to have a fixed home. Deceased then said, "Oh, no. I have a fixed home." Witness understood him to say that his niece was or had been his housekeeper. Deceased said, his fixed home was at Felton, on the Darling Downs, a bit of the most beautiful country that was to be found not only in Queensland, but in the whole of the Australian colonies. He emphasised the remark about it being his fixed home. Deceased further stated that it took him about one-third of the year inspecting his stations, and the remainder of the year he stayed quietly at Felton. Witness wanted to make inquiries about station life on behalf of his son, and deceased remarked, " Just put all your inquiries in a letter and address it to my home at Felton and I will answer it." Witness had many general conversations with deceased, but the latter made no further reference to his home. Witness when he met deceased he used to address him as Mr Tyson, and deceased told witness to call him " Mr. Smith." (Laughter.)

Cross-examined by Mr. Lilley: Deceased appeared to be getting deaf in 1896. Deceased told witness that in travelling he went to New South Wales, Gippsland in Victoria, and to many parts of Queensland. He did not tell witness that all his relatives were in New South Wales. The only private matter he spoke of was the first £60 he had saved, and which, he said, took him three years to accumulate. Witness did not remember having heard the name of Felton before deceased mentioned it. Deceased never used the expression that "he was always on the wing," or that he was a wanderer.

Brisbane Courier, 9 November 1900 P 7.

For the convenience of the witness, who wished to return to Rockhampton, the evidence of Mr. Frank Alex. Huet was interposed during the reading of the late Mr. Tyson's diaries. Mr. Huet deposed that he was a dental surgeon,' practising at Rockhampton. He first met Mr. Tyson at Toowoomba in 1892. He thought him a most eccentric man. Tyson had said he would rather sleep out in the open than in the best of beds. " How about your toilet?" queried the witness. " My what?" said Tyson. " Oh, soap, towels, &c," replied the witness. " I have never used soap in my life," re-joined Tyson decisively. Witness subsequently met Tyson frequently, in the course of his travels. On one occasion at Emerald, in 1896, he had a long talk with Tyson. Travelling many thousands of miles, during the year, and being a bachelor, I presume," said witness, " that you have no fixed home." " Why do you ask?" queried Tyson. Witness answered that he considered it would be an advantage to a bachelor not to have any fixed home. " Oh, no," re-joined the millionaire, " I have a fixed home at Felton, on the Darling Downs-a bit of the most beautiful country to be found not only in Queensland but in the whole of the colonies." Witness under- stood him to add that his niece was, or had been, his housekeeper.

Mr. Feez: Did he say anything about inspecting his stations?

Mr. Huet: Yes; he said he got through his work in about a third of the year, and that he remained quietly at Felton for the remainder.

Witness related several anecdotes. " How do you do, Mr. Tyson?" he said to the deceased in some public place one day. " In future," said Tyson, sotto voce, " call me Smith." One day Tyson assured him he never wore a starched shirt in his life. " Why do you wear that collar, then?" observed witness. " It makes one look tidy a bit," was the answer. " But a man of your means can afford to be indifferent in matters of dress. If I disliked wearing starched shirts, I wouldn't wear starched collars." " Do you know what this collar is?" said Tyson quietly. Witness looked at it closely. " It's celluloid." " Yes," replied Tyson, " I've been wearing it for eighteen months."

Witness: Walter C Hume, Member Land Court, formerly District Land Commissioner
Brisbane Courier, **16 November 1900, P 7.**

Witness identified official sale lists of land to Tyson dated 9th March, 1876, and 11th April, 1877, in which his residence was described as Felton, together with a declaration of fulfilment of conditions on Tyson's selection No. 392 and witness's certificate that the conditions had been fulfilled. The conditions required that the selector should reside on the selection, or on the adjoining freehold. Tyson did the latter, and the certificate set out that circumstance, that he had so resided for six years prior to the date of the certificate. In 1879 witness, being under the impression that men were taking up land for Tyson, held an inquiry into the bona-fides of a selector, and Tyson thereupon made a declaration, which he sent to the Minister for Lands, and in which he repudiated the allegations made against him, and set out that " during his many years' residence in Queensland" he had never been a party to the violation of the land laws of the colony. Witness had no general conversations with Tyson, either about private or public affairs. His actions in regard to the selection just mentioned caused Tyson to look coldly upon him.

Witness: John Kelly, Railway guard, Southern Line
Telegraph, **20 November 1900, P 2.**

Witness stated that deceased had frequently travelled on the train with him between Cambooya and Wallangarra and from Brisbane to Cambooya. Deceased on several occasions, gave witness letters, after he had opened them, and asked witness if he would send them home for him." Witness used to give them to the station-master at Cambooya! Witness knew that deceased travelled under the name of Smith. Witness had put a box labelled Smith out at Warwick, and on getting to Wallangarra deceased asked for it. Witness denied that deceased had a box on the train, and deceased described it, when witness told him that it had been put out at Warwick; Witness asked the reason for addressing it " Smith" and not "Tyson." He replied, " When I leave home and go down south one man wants to know what sheep are good for this part of the country, another wants to know, what sheep are good for that part of the country and I perhaps do not know as much about a sheep as you do. That is the reason why I travel under the name of Smith." Witness met deceased on one occasion on the Hawkesbury River, when deceased said he was going to Gippsland in Victoria. Witness, said, " I suppose you will not be home for some time?" Deceased replied, " No; I will not be home for some time."

Witness: Sir John Lackey, President Legislative Council NSW
Brisbane Courier, **24 November 1900 P 2.**

Deposed that he induced Tyson to become a member of the Warrigal Club in Sydney, and that Tyson seemed favourable to a suggestion of his that he should give up wandering and make the club his home.

Witness: Donald Mackintosh, M L C, Cambooya
Telegraph, **20 November 1900, P 2.**

Witness stated that on one occasion he accepted an invitation from deceased to visit Felton. Witness was accompanied by deceased to the house. When they got to the gate witness wanted to dismount to open it, remarking, that he was the younger man. Mr. Tyson, however, insisted on opening the gate, replying, " You are coming home with me as my guest." In the morning deceased showed witness a tree and said, " I took that tree from my mother's garden in New South Wales after her death and planted it here at my home at Felton in order to serve to remind me of her." The fact that he should so remember his mother impressed itself on witness's mind at the time. Deceased was very proud of Felton, and often spoke of it as the garden of Queensland. Deceased gave witness his address as at Felton. He interested himself a good deal in matters pertaining to the Jondaryan Divisional Board.

Cross-examined by Mr. Lilley: Witness remembered the tree incident because at the time he was struck with admiration for deceased who, it was generally considered, did not think of anything but making money. Deceased did not care much for women of the "lah-de-dah" sort, but if he saw children with clean and neatly patched clothes he would remark, "There is a good mother." Deceased was peculiar, and would not answer a direct question. If a person asked him how he made his fortune he would reply, " Well, my man, I made it by minding my own business." Deceased often told witness that he was glad to get back to Felton.

Witness: John McDonald, Manager, Tinnenburra
Telegraph, **24 November 1900, P 2.**

He stated that his wife's name was Elizabeth Hewitt, and she was a niece of the deceased, and entitled to a share in the estate. Her share would be about the one sixty-fourth part. He formerly managed Maranoa, in New South Wales, which adjoined Tinnenburra, and was sold after the death of the deceased. Both stations were under the one management. Deceased described them as his Warrego stations. Witness went to Tinnenburra in 1878, and had been there ever since. First met deceased in 1876. After meeting deceased went to Felton to see him about going to the Warrego for horses. Witness went for the horses, and returned to the Downs again. On returning witness was employed at Clifton for 12 months, after which he went to Tinnenburra. Witness was married in January, 1878, and took over the management of Tinnenburra in March of that year. His wife had been living at Felton for about three years before her marriage. After 1878 witness did not see deceased again until 1883. From that year his visits to the Warrego were made on an average about every two years. In 1890, when they were boring for artesian water, deceased twice visited the station. Witness often saw correspondence at Tinnenburra addressed to the deceased. Letters were usually addressed "James Tyson, Tinnenburra." Witness used, to forward the letters to Felton or Sydney, wherever witness thought deceased was to be found.

Deceased subsequently gave witness instructions to open all letters, and if there were any relating to matters that he could not deal with to forward them to him (deceased). Letters came from all parts of the world—mostly begging letters. Deceased never gave him any instructions about forwarding the letters to any particular place. He frequently sent witness a telegram as to where he was going, whether to the north or south. Witness accepted the telegrams as instruction. When witness did not know the movements, of the deceased, he sent the letters to Felton. Deceased told witness in 1892 that he found the climate of the Downs too cold for him, and that he required to avoid it in the winter. Deceased said, "I may go to the Lachlan, or I might put in a month or so with you on the Tinnenburra." Witness's experience of deceased was that he would not stop in any place once his business was concluded. He often told witness so, and said, " I am never as well as when I am travelling." Deceased left clothes at Tinnenburra on two occasions. Once he remarked, "I have reduced my swag; I have left some clothes behind." Deceased had spoken to him about exchanging Felton for land on Tinnenburra and mentioned that he would require four acres to one. Witness met deceased in Sydney about the year 1886, when he said, " I have had a vault built at Campbelltown, and I have had my father's and mother's remains removed to it." He further stated that he intended to be interred in the vault. Deceased had told witness that "he was a wanderer, and always travelling." By that witness understood him to mean that he was much harder worked than any of his managers. Deceased often spoke of his New South Wales properties, and seemed very fond of the Lachlan (River). He described it "as a fine place to live, and there were plenty of wild flowers." Referring to the place where his mother had died, he told witness that the old house had been burnt down, and it was his intention to have it rebuilt. In 1896 deceased told witness that he was negotiating for the purchase of a large property in Sydney, to cost about £8,000, and that he went to Sydney to conclude the sale. He, however, had been advised that Sydney property values would fall lower and had not affected the purchase. Witness brought some pear, quince, and cherry trees to Tinnenburra from Wilton, New South Wales. These were obtained from the garden adjoining the place where deceased's mother died.

Telegraph, **27 November 1900, P 3.**

John McDonald, manager of Tinnenburra station, and husband of a niece of Tyson's, was cross-examined by Mr. Feez.

Counsel: You knew from Mr. Tyson's letters that Felton was the starting point and ending point of all his journeys?

Witness: I knew Felton was one of his principal addresses.

His Honour (to witness): Answer the question put to you.

Witness replied to counsel's question in the affirmative.

When Tyson engaged you in September, 1876, he described himself in the agreement as of Felton? - Yes.

Your wife was housekeeper at Felton before you married her? - Yes.

For how long? - Three years.

Whose housekeeper, was she? - I don't know.

His Honour (surprised): You don't know whose housekeeper she was? - I do not.

Do you know who paid her wages? - I presume Mr. Tyson.

And you ask me to believe you don't know whose housekeeper she was? - I can't say whose housekeeper she was.

Mr. Feez: Whose do you think she was? - I presume Mr. Tyson's. I don't know who paid her, or what she was getting.

She never told you? - Never.

Were your relations such that she never talked to you of her post? I don't remember her telling me anything about it.

You married her from Felton? – Yes, her brother was living in the house at the time.

His Honour: What was he doing? - He was overseer.

I suppose you sometimes visited her at the station? - Yes.

Often? - Not very often.

Mr. Feez: There was another James Tyson? - Yes.

How did you describe him? - James Tyson, junior.

Of Tupra? - Yes.

As distinguished from James Tyson, of Felton? - Yes.

You say that Tyson told you he wished to be buried in the vault at Campbelltown? - Yes.

And you know that other witnesses say he told them the same thing? - Yes.

Don't you think it remarkable, then, that all you people who got money out of him did not take the trouble to remove his remains there in accordance with his wish? We have not got all the money yet. A circular has been sent round among the relatives, and all of them have consented to pay their share of the expense of removing his remains to the vault.

When was that circular sent out? - A few months back.

Since this action was begun? - Possibly six months ago.

The remains of Tyson's brother who died on the Warrego were not removed to the vault? - No.

Telegraph, 27 Nov 1900, P 3.

Mr. John M'Donald, manager of Tinnenburra station, who was the first witness called on behalf of the defendants, cross-examined by Mr. Feez, stated that as a rule the letters he received from deceased were addressed from Felton. Witness did not forward begging letters to the deceased but destroyed them. Deceased sometimes spoke of visiting his stations and returning to Felton. In his letters sent to witness from Felton deceased often used the expression " I have just returned." Witness knew that Felton was deceased's starting and ending point. In the agreement made between deceased and witness in 1876 about going to the Warrego for horses deceased described himself as "James Tyson, Felton." His wife had been housekeeping at Felton, but he did not know whose housekeeper she was.

His honour: Do you expect me to believe that?

Witness: I do not know whose housekeeper she was or who paid her. I presume Mr. Tyson paid her.

Mr. Feez: Are you at such distant relations with your wife that you mean to say that she never told you?

Witness: I do not know as a fact whom she was employed by, or who paid her.

Mr. Feez: Yon. married her from Felton, and you do not know whose housekeeper she was?

Witness: Her brother was living in the house as well as deceased.

By Mr. Feez: Deceased did not leave the impression on witness's mind that Felton was his home. Witness considered that he had no home, and that he lived at Felton merely because it was convenient for him to go from there to his other stations. Witness's wife told him that some fruit trees were brought to Felton by Mr. Tyson Doneley from the garden of her grandmother in New South Wales.

Witness: Clarence McIvor, Stock expert, Sydney
Telegraph, **28 November 1900, P 2.**

Witness stated his headquarters were in Sydney. Had travelled all over Australia. Had met deceased at Narrandi, in Sydney, and also in Melbourne. Met him frequently up to the year 1897. Saw him at Felton in February 1897. Deceased had given witness an invitation to Visit him if ever Witness was near any of his properties. Witness stayed a week at Felton and discussed general subjects with deceased. Witness had said to deceased, "The time is gone for the Darling Downs to carry merino sheep it is a great pity that you own so much of it, as it will keep it from carrying a large population." Deceased replied, "Oh, that is where you make a mistake, you are like a lot of other people who jump at conclusions. I am ready and willing to exchange my Darling Downs property for land out back, or to sell it right out." Deceased said the price he wanted was £3 10s. per acre for Mount Russell and £2 19s. for Wybie. Deceased further said he would be able to leave Felton in two hours. He explained how he came by the properties. Witness asked if he could make it public that deceased was willing to part with the Darling Downs properties and deceased replied " Oh, yes." Deceased told witness that he was getting old, and that he would go back to the place where he was born. To use his own language, deceased said, " I want to go back and be buried alongside my mother." During a conversation in 1895 witness had told deceased that practically he had no fixed abode, that he was compelled to travel, and deceased had replied, " You are like me then, I am always travelling." Witness said, "I thought you lived at Felton?". Deceased replied, "No; Felton is only my principal Queensland address." On the visit in 1897 witness had made a remark about there being no garden at Felton. Deceased replied, "I have no time and I did not intend to stay here so long." Witness said, "Who built this house?" ' Deceased replied that he did and witness re-joined, "Well, did you not intend to make a home here when you built it?" Deceased replied to the effect that it was no use offering land for sale, unless there was a good house on it, as buyers thought more of the house than the land. On the. subject of a will witness had said it was a good thing in some cases when persons died without a will. Deceased asked witness if he really thought it was a good thing to die without a will. Witness replied, " Yes, sometimes, of course, there is a much heavier duty." In reply to a question put by witness as to whether deceased had his affairs in order, deceased said "No; there was a will made some years ago." He then went into the house and produced what witness took to be a draft copy of a will. Deceased gave as a reason for not making a will that he had no time, that he was always travelling, and that his properties were scattered over three colonies. He further said he hoped to consolidate them, and then remarked, " Mr. Macdonald-Paterson, my solicitor, has gone home; I will have to wait till he comes back." Witness pointed out to him that he should have a will dealing with his properties in the three colonies, and explained to him how Queensland could claim duty on all his properties if Queensland was decided to be his home. Deceased said, "That is wrong, my home is not here. I have come here on business. My home is where I was born, and where I intend to be buried. New South Wales is my colony; this is Buchanan's home. I have a room on every one of my properties. Heyfield is as much my home as Felton is." On another occasion deceased had approached witness about investing money in city properties in Sydney.

Cross-examined by Mr. Woolcock: Witness had noted down a lot of Mr. Tyson's utterances. "The books were now in South Africa, having been taken there when witness enlisted in the Imperial service at the beginning of the year. Witness joined in Sydney as a special service officer. Witness was sure he met deceased in Sydney in June, 1895.

Mr. Woolcock: Would you be surprised to learn that deceased's diary shows that he was on his way to Meteor Downs at that time?

Witness: No.

Mr. Woolcock: Would you pit your memory against his diary?

Witness: I know something about the diary. Deceased used to enter plans a week ahead and not carry them out.

Telegraph, 28 November 1900, P 5.

Mr. C. M'Ivor, further cross-examined by Mr. Woolcock, stated that on one occasion at Rockhampton in 1897, the deceased was marking entries in his diary and he told witness he was making his arrangements for the week. Witness knew his arrangements were not carried out, as he said he. would be back from Meteor Downs on June 12, and he was' not. Witness would not be surprised to hear that there was no reference to June 12 in the diary. When witness was staying at Felton deceased "plotted out" a visit to Mount Russell, but then suddenly turned round and said, "We will not go." Witness studied Mr. Tyson very closely, and often saw him making entries in his diaries which were not carried out. Witness knows this from the instance at Rockhampton and others. Witness was at Felton from February 17 to February 24. Deceased was there the whole of the time. If the diary showed that deceased was there only two days, he would say it was inaccurate.

Telegraph, 28 November 1900, P 1.

In further cross-examination by Mr. Woolcock, Mr. M'Ivor said that when deceased told him (witness) that Felton was his principal Queensland address, he also said, "I have more veneration for Heyfield." Witness was interviewed by a representative of the Telegraph on his return from Felton in 1897. Witness had not gone to Mount Russell or any of the other stations. He went out for a few rides with Mr. Buchanan.

Mr. Woolcock: In the interview you say that you have just returned from spending a week with Mr. Tyson in going over his Darling Downs properties. Is that correct?

Witness: The interview is fairly correct for a newspaper article.

Mr. Woolcock: Is the interview correct?

Witness: Journalistic experience and what you require in court are quite different things.

Mr. Woolcock: What do you mean by that?

Witness: The gentleman who wields the blue pencil (the sub-editor) adjusts according to his own idea.

Mr. Woolcock: You did not spend a week going over the Darling Downs properties?

Witness: I think I told the reporter that I had just returned from spending a week at Felton.

Continuing, witness repeated details of conversations he had had with deceased about his will and his home. He added that deceased had told him that he had an Invitation to go with Sir Hugh Nelson to England about jubilee time. Witness advised him to go, and deceased replied, "I would be taken amongst a lot of swells, and I would not be happy. I would dearly love to see the Queen." Witness told him that if he went with Sir Hugh Nelson, he would be almost sure to see the Queen, that he would he presented at court, and that he would be well treated. Witness showed him a picture of a court suit in a magazine, and said to him, " Before you could be presented to the Queen you would have to get a suit like this," Deceased replied, "I would not put that on if they gave me England."

Being further questioned about the will and what deceased had said about his properties, His Honour asked witness how it was that deceased took him (witness) so much into his confidence. It was notorious that deceased rarely confided private or business affairs to anyone.

Witness: Because he found out early that he could trust me. Besides, I used to contradict him, and he said it was refreshing to have someone who would do that.

By Mr. Woolcock: Witness did not volunteer to give evidence in the ease. Mr. Slattery, solicitor for one of the beneficiaries, in Sydney, knew that he could tell them something about Mr. Tyson, and he had made a remark one day to that gentleman's clerk that deceased had not lived at Felton.

In reply to a question about a certain hotel bill, witness stated that he had forwarded the amount by a friend to be paid.

Mr. Woolcock: Would you be surprised to hear that a summons has been issued against you for that amount.

Witness: I suppose you gentlemen have stirred it up. I have lived too long in the world to be surprised at anything.

In answer to his honour witness stated that after starting for South Africa he was injured in a storm, and as a result he had lost his left leg. He did not get further than Albany. His box of books went on to South Africa, the vessel being wrecked at Port Elizabeth.

Witness: Edmund Morey, Police Magistrate, Maryborough
Brisbane Courier, **16 November 1900, P 7.**

Witness detailed a conversation he had with Tyson at Rockhampton in October, 1895. It related to the latter's private affairs. Tyson, in the course of it, said he could not trust anybody, and showed witness a. pocket-book, in which he recorded all his transactions. When he filled one book, he began another, and the old books he numbered and put away carefully at home. Witness, knowing well enough where his home was, did not think it necessary to put a question on the matter.

***Telegraph*, 15 November 1900, P 5, In: SECOND EDITION.**

Witness stated that he met the late Mr. Tyson in May, 1846, in the reed beds of the Lachlan River, at the junction with the Murrumbidgee River. Since then, had met him in Queensland. Remembered him going to the Darling Downs. Witness last saw him in Rockhampton about October, 1895. Deceased told witness that he had been out west, and he found the outside work getting too much for him. Witness suggested that he should appoint someone to relieve him of it. Deceased shook his head, and said, "No one man could do it." "Witness said. "Perhaps two could?" Deceased replied, "I cannot trust anybody." Witness re-joined, "That is just your weakness, Tyson, you cannot trust anyone." Witness gave deceased some wholesome advice, and in the course of further conversation asked him how he kept touch of his affairs. Deceased drew out a small memo book and handing it to witness said, " Look at this." Witness read an entry in the book and then remarked, "But these books are soon filled up!" Deceased replied, " As each book is filled up, I put it away carefully with the rest at home." Witness assumed that in using the word home deceased was referring to Felton. Witness wrote the address of his eldest son in the deceased's pocketbook at the request of the deceased.

Witness: George S Murphy, accountant
***Telegraph*, 16 November 1900, P 2.**

Witness stated that he was formerly a director of the Q.N. Bank and also a member of the old Union Club. Deceased was a member of the club and used to stay there. Often had chats with the deceased at the club on Sunday mornings, when deceased used to show witness his letters. Deceased used to refer to his food and complained that he could not get corn beef, while at home he always had corn beef. Witness asked him why he did not, when he was dissatisfied with the accommodation, make his home in Brisbane. Deceased replied, " I will never do that, Felton is my home." The conversation took place about 1890. Deceased always gave witness to understand that Felton was his home.

Witness: Sir Hugh Nelson KGMG, MLC, Toowoomba
Telegraph, 15 November 1900, P 7.

Sir Hugh Muir Nelson, K.G.M.G., President of the Legislative Council of Queensland, stated that he had been Prime Minister and Treasurer of the Legislative Assembly of Queensland. Had lived continuously in the Toowoomba district for a period of 40 years. Knew the late Mr. Tyson intimately for 20 or 21 years prior to his death. Deceased often came to see witness at his rooms at Parliament House. Deceased frequently went away to visit his properties in the other colonies. Latterly, since 1893, deceased never went away without telling witness that he was going away from home. Deceased was a member of the Legislative Council, and was sworn in in the year 1893. Deceased attended to his duties in the Legislative Council with fair regularity. It was lest his absence should be missed that he used to come to witness, and if witness disapproved, he would put the journey off. Deceased always referred to Felton as his headquarters, and spoke of coming back. Witness never asked him in direct terms where his home was. It was generally understood. Deceased would say, "I am going home on Friday morning, " and witness would reply, " I am going home too," and they would go together. Deceased would say if he was 'going to the Central districts, "I will be home again in three weeks." Witness had written to deceased, and always addressed the letters to Felton.

Witness wrote to him in 1879 regarding the election for Northern Downs, asking him for his support. Witness then thought deceased a man of local interest. Witness was not aware if ever deceased contemplated becoming a candidate for Parliamentary honours. Deceased purchased £200,000 worth of Queensland Government Treasury bills, which were; issued in 1891, before witness became Treasurer. Witness offered him the money before the bills matured, but deceased would not take it. He did not wish to do so even when the bills fell due, and afterwards on witness's advice he purchased Government Savings Bank stock.

Mr. Lilley: That was a good investment of his with the Government taking up the Treasury bills.

Witness: I dare say.

His Honour: Deceased's notion was regarded as a patriotic one at the time

. Mr. Lilley: Deceased was helping the country and himself at the same time. In reply to further questions by Mr. Lilley, Sir Hugh stated that deceased often spoke of his properties in New South Wales. He never said anything about his family relations. Deceased never spoke to witness about exchanging Felton for western land, or that he contemplated buying a house in Sydney.

Brisbane Courier, 16 November 1900, P 7.

Deposed that he had known Tyson for twenty years prior to his death. After 1893, when Tyson was sworn in as a member of the Legislative Council, he never went away without telling witness that he was going. If witness disapproved, he would put the journey off. Deceased always referred to Felton as his headquarters, and when witness wrote to him, he addressed his letters there. Witness had never asked in direct terms where his home was. He understood it to be Felton. Deceased would say, " I am going home on Friday morning," " I am going to the Central district, and will be back in three-weeks," &c. When in town, Tyson had a habit of visiting him at his house in the morning, between 6.30 and 7 o'clock. In 1891 Tyson took £200,000 worth of Treasury bills, and when they fell due, he declined to take the money. On being informed that he must he then invested it in Savings Bank stock.

Mr. Lilley: That was a good investment for Tyson, was it not? What interest did he get on the Treasury bills?

Witness: 4½ per cent.

His Honour: I know it was considered at that time a patriotic action on Mr. Tyson's part.

Mr. Lilley: With people who did not know the rate of interest. (To witness): What was the interest on the Savings Bank stock?

Witness: 3½ per cent.

Witness: Mrs Maria Agnes Parsons, Housekeeper Heyfield
Telegraph, 24 November 1900, P 2.

Witness stated that she had been housekeeper for the deceased at Heyfield from 1893 until the time of his death. Deceased frequently visited Heyfield. In 1897 she believed he was there about six times. his visits last three, four, and six days. On one occasion she said to him, " I suppose, Mr. Tyson, you have many homes" He replied, " Yes, I feel more comfortable at Heyfield than on any of my other stations." He only mentioned Felton to her once; that would be about the year 1890. He told her that Mr. Buchanan managed Felton for him, but he did not compare that gentleman with other station managers.

Witness: William Henry Passmore, Stationmaster, Stanthorpe
Brisbane Courier, 16 November 1900, P 7.

Witness stated that he acted as accountant and secretary for Tyson from October, 1880, to March, 1882. He lived at Felton. Important letters were written by Tyson on a slate, from which witness copied them. All account sales and bank passbooks from New South Wales and Victoria, as well as Queensland, were forwarded to Felton. When Tyson went away, he usually said he was " going on business." ' Afterwards when passing through Stanthorpe Tyson always got out of the train to speak to witness. Returning from the South he used to speak of " going home." At Felton witness had heard him say he was " glad to be back again." Correspondence arriving there in Tyson's absence had to await his return. Prior to the building of the new house Tyson had spoken of building a house " worthy of him and an ornament to the place."

Witness: H H Peck, Stock Agent
Telegraph, 24 November 1900, P 2.

Witness deposed that he had stayed at Heyfield with deceased in 1896. Noticed deceased's wardrobe and made a remark, "That if Tyson had a similar wardrobe at each of his stations as to that he had there, he must have as many clothes as any person in Australia." Deceased in speaking to him referred to Mr. Buchanan as his right-hand man, and looked on that gentleman as a sort of inspector-general to supervise where he (Mr. Tyson) could not.

The Chief Justice: Is that correct?

Mr. Feez: There is no other evidence of it, your honour. The witness is the only one who says so. Deceased sent Mr. Buchanan to Heyfield when Mr. Mills died, but that is all."

Witness: Hon Robert Philp, Premier of Queensland
Telegraph, 17 November 1900, P 2.

Witness stated that he knew the late Mr. Tyson from the time deceased foreclosed on Felton station in 1870. Had many conversations with him. Deceased proposed to go to the old country with Sir Hugh Nelson in 1897, and at that time he asked witness if witness would consent to Mr. M'Donald, of Tinnenburra station, acting as his attorney. Deceased always spoke of Felton as his headquarters, and all communications were addressed to him there. Witness had done business for deceased in connection with the New South Wales stations, and copies of all invoices, &c., were sent by his direction to Felton. All account sales of wool and cattle disposed of in New South Wales and Victoria were sent to Felton. At the time witness was managing director of Messrs. Burns. Philp, and Co. Deceased always considered Queensland was the best colony of the group, and told witness frequently that he was gradually withdrawing investments in the other colonies with the view to placing them here. In discussing selling Felton to the Government under the Agricultural Land Purchase

Act he stated "that he would not care about selling the place as it was his home." He was alluding to the place as a whole. Deceased said he enjoyed the best health in Queensland. When deceased returned from New South Wales or Victoria, he always spoke of going home to Felton. That was the only permanent residence witness knew deceased to have. Deceased made an application to be placed on the North Brisbane electorate as a freeholder. Witness was present when the claim was signed.

Cross-examined by Mr. Lilley: Witness had met deceased outside of Queensland. About one-half of deceased's estate was invested in Queensland. When he came to Queensland all deceased owned was Tinnenburra, and he had a mortgage on Felton. In the last budget speech witness had forecasted getting about £70,000 as succession duty out of the deceased's estate.

Witness: James Porter, Grazier, Worth Grass
Telegraph, 15 November 1900, P 7.

Witness, residing at North Branch, within six miles of Felton, stated that he had known deceased from 1870. Saw him frequently at Felton. In conversations deceased always spoke of Felton as his home. Witness remembered Mr. Hogarth's body being discovered in the creek. Witness was then working at making the coffin when deceased remarked, "I have just been wondering who would do that good office for you or me." Witness replied, I will get a coffin made, but you are so much of a wanderer that the black boy will find you dead in camp some morning." Deceased said, "Well I have given up those lonely journeys now; I am getting old. If I only take a trip to Brisbane, I am longing all the time to go back home to Felton to rest." About 12 months before he died, witness saw him going to the Pittsworth land sale, and deceased had said he felt so ill that he thought of turning back home.

Witness: William Roberts, Drover, Ipswich

Telegraph, **17 November 1900, P 2.**

Witness stated he met the deceased Mr. Tyson in the "fifties." Meeting him again in 1872 Tyson told him that he had bought Felton. Witness remarked, "So you are going to become a banana man." Deceased replied " Yes, I am going to make my tourai (a native name for home) here." Queenslanders were termed " bananalanders." Witness saw him again in Ipswich, when deceased said he was going to New South Wales, and it would be some time before he would be back home. Witness understood him in referring to home to mean Felton.

Witness: Mrs Frances Sheil, half-sister
Brisbane, **23 November 1900, P 2.**

Witness stated that Tyson told her husband to have the vault in the Campbelltown Cemetery made to hold six bodies, those of his (Tyson's) parents, his three brothers, and his own. Tyson often spoke of being buried there himself. The deceased, she said, had a home at Ivanbury (N.S.W.)

Isabella Sheil, daughter of the last deponent, gave corroborative evidence.

Witness: Arthur B Stewart, Coal Merchant, S Brisbane
Telegraph, **17 Nov 1900, P 2.**

Witness stated that in 1880 he was a steward on the steamer Governor Blackall, running between Sydney and Brisbane. Remembered deceased travelling in the steamer. Deceased did not travel under his own name, but under the names of Mr. Smith, Mr. Walker, and Mr. Stokes. The assumed name used to cause confusion, because they did not provide a berth for him, as the name of Mr. Tyson did not appear in the passenger list, and they could not find Mr. Smith who had the berth engaged. (Laughter.) Witness afterwards joined the Katoomba. On one occasion when coming from Sydney witness was speaking to the chief engineer when deceased came up to them. The chief engineer said, " Are you going from home again. Mr. Tyson? "

Deceased said, "Do not call me Tyson, my name is Walker." The engineer said, "That is bosh; why do you not want your name to known?" Deceased replied, " I do not want my name telegraphed to Brisbane before I get there." The chief engineer said, "Are you leaving home, Mr. Tyson?" He said, "No, I am going home; my home is on the Darling Downs." The chief engineer re-joined, "I always understood your home was in Victoria." Deceased answered, "No, my home is on the Darling Downs." Witness subsequently kept a hotel at Pittsworth. Remembered Mr. Daniels being elected for Cambooya. Deceased was in Pittsworth on the election day, having driven in in a buckboard buggy with two horses. On that occasion witness had a conversation with him about staying in Pittsworth, and witness said he hoped deceased would lend Mr. Daniels a hand. Deceased said "No; I scratched my pencil through all the names, so I make a friend or foe of no man." He also declined to stay, stating that he preferred to go home to Felton.

Cross-examined by Mr. Lilley: Remembered the conversation on the Katoomba through an incident that occurred. Deceased invited the chief engineer to have a drink. The engineer consented and asked for a " ginger ale." Deceased decided to have a ginger ale also, but considered the one bottle enough for both, and thought they should "split" it. (Laughter.) The engineer replied in rather a resentful way, that "he (Tyson) could have the lot." Witness then served the ginger ale to deceased, and charged 1s. for two drinks. Deceased protested that 6d. was the proper charge, and eventually reported the matter to the captain. (Laughter.)

Witness: Joseph Stirling JP, Blacksmith, Childers
Telegraph **13 November 1900, P 2, SECOND EDITION.**

Witness stated that he had formerly resided at Toowoomba for a period of 30 years. He left Toowoomba in 1895. Knew the deceased intimately. Had done work for him, and frequently had conversations with him. Deceased had stated on one occasion in reply to a question put by witness, "Felton station is my home." He further said, "You have not been up since I built my new home." Deceased gave witness an invitation to go and see him. Witness did go to the house, but deceased was not there. Witness supplied the cedar, which deceased selected. He would not take anything with a flaw in it, and said he would pay the best price for the timber. He told witness that he had selected every stick of timber that was in his house. Witness had made the ironwork for the house at Felton. Everything had to be made extra strong. When witness went to Felton the only human person, he saw was the cook. The men were working some miles away. He spoke to deceased afterwards, and the latter remarked, "I keep nothing on any of my stations that does not earn its own rations. Every dog has its own beat, and if it is off its own beat Tyson knows why." The house was splendidly furnished and was well ventilated. Deceased was very proud of Felton station. Witness considered it was the pick of the Darling Downs.

Cross-examined by Mr. Lilley: Deceased did not tell witness that he had a house at Tupra in New South Wales. He rarely discussed anything relating to family mutters. Witness did not know whether the deceased had everything made especially strong for his other stations. Deceased prided himself on being a bit of a blacksmith. He once discussed with witness about entering witness's employment, and asked the amount of wages he would be likely to make. Witness told him that he would perhaps be able to earn about 30s. per week, and he replied, "That would hardly pay me." Witness never met deceased anywhere except at Brisbane, Toowoomba, and Felton.

***Brisbane Courier* 13 November 1900, P 2.**
Witness deposed that he had known Tyson very well during the thirty years he (Stirling) had lived at Toowoomba. Tyson told him that of all his stations he looked upon Felton as his home. Witness by request once visited the station. Tyson was absent, and the only living things he found on the whole homestead were a girl, an old cook of 80, and a cat. He was not surprised later on to learn that the girl had drowned herself somewhere. The place was enough to send one

melancholy mad. When he mentioned his visit to Tyson, the latter said gleefully, " Oh! I keep nothing on my establishments that does not earn its rations. Every dog has its own beat, and if it is not there James Tyson must know 'the reason why."

By Mr. Lilley: He had frequently succeeded in " bleeding" Tyson for subscriptions to military tournaments, churches, schools, &c. Tyson would never allow his name to appear. The donations were always from " A Friend."

Witness: James Tyson II, Retired grazier, Near Hay
Telegraph, **27 November 1900, P 3.**

Witness stated he was a nephew of the deceased. Witness lived with deceased on the Lachlan station from the year 1848 to 1850. Afterwards went journeys with him. Remembered deceased acquiring Juanbung, Deniliquin, and Tupra stations. Witness managed Tupra for deceased up to the year 1892. Witness's brother managed Juanbung up to his death in 1878, when Juanbung. and Tupra were worked as one holding. When deceased sold Deniliquin he lived at Juanbung until he acquired Heyfield in Victoria. Deceased built a house at Heyfield and lived partly there and partly at Juanbung in 1865 and 1866, In the latter year he built another house at Juanbung. The houses at Heyfield and Juanbung were superior to the house built on Felton. After building the house on Juanbung deceased lived there until 1870, when he took possession of Felton. Witness used afterwards to see deceased at Juanbung, and. Tupra. Witness received letters for deceased addressed to Tupra.

Mr. Lilley: You were general manager for deceased over all his New South Wales stations?

Witness: Yes

Mr. Lilley: Did deceased keep an eye over you?

Witness: Oh, my word. He kept in close touch with all his properties. Continuing, witness said deceased did not notify him of all his (deceased's) movements. When he was about to visit a station where witness was, he generally acquainted witness of his intention.

Deceased must have been travelling the greater part of the year. Deceased kept up the house at Juanbung after going to Felton. Witness's brother lived in the house from the time it was built until 1877. Deceased had a library at Juanbung, consisting of about 100 books. He kept a box in his bedroom for his private papers. Deceased never managed a station himself. Deceased frequently told witness that he had better health travelling about than stopping at any one place. Witness believed that he travelled, more during the years just before his death than formerly.

His Honour: Who became his general manager after you left him in 1892?

Witness: I believe he generally supervised matters himself.

His Honour: Why did you leave him?

Witness: I was otherwise engaged for a year or two, and I did not go back. Besides deceased wanted me to come and live in Queensland.

Mr. Lilley: You would not come over here?

Witness (shaking his head): No. I would not care to live here. It might be all right to come here and make money, but I would not live here.

Mr. Lilley: You had means of your own when you left deceased?

Witness: Yes; I had an independence of my own.

By Mr. Lilley: Felton was of more value per acre than Tupra. Deceased had spoken to witness of exchanging Felton with the Government for land on the Warrego. Witness overheard deceased call Felton his home. Often heard people ask deceased where he lived, and he always put them off with the reply. " O, I live everywhere."

The court adjourned until the afternoon;

On resuming,

The witness James Tyson, continuing his evidence, stated that deceased had objected to the sale of his " old home " in New South Wales, and it had not been disposed of. Deceased visited his mother once a year before she died. He had told witness that he was "a wanderer and always on the wing." Deceased did not sell any of his New South Wales' properties after disposing of Deniliquin in 1861.

Cross-examined by Mr. Feez: Deceased never told witness that he intended to be buried in the vault at Campbelltown. Witness did not think deceased was more intimate with him than with other relations. Witness believed he visited Tupra, after 1870, on an average about once a year. His stays extended for a few days, or two or three weeks. Witness believed he was at Tupra in the years 1872, 1873, and 1874. Witness in writing to deceased in 1874 used the expression, " As you have not been here for a long time." Witness probably meant a year by that expression.

Mr. Feez: Why did you leave your uncle's employ?

Witness: I decline to answer; it has nothing to do with this case.

Mr. Feez: You must answer; it has everything to do with the case.

Witness: I was taken into the Divorce Court.

Mr. Feez: You were taken in?

Witness: I was. not taken in, I took the woman in and that is the same thing.

Mr. Feez: That is a very different thing. As a matter of fact, you asked for a nullity of your marriage with the woman you had been living with as your wife for many years?

Witness: Yes.

Mr. Feez: And you swore that you did not know you were being married to her when the marriage ceremony was performed.

Witness: That has nothing to do with this case.

Mr. Feez: It has a great deal to do with it. Did you not swear that you thought that the priest was giving her some sort of a sacrament or remission and not performing the marriage ceremony! Witness: I do not think I need answer.

His Honour: You are asked if you knew whether the marriage ceremony was being performed.

Mr; Feez: Did you not swear that you did not know the ceremony of marriage was being performed!

Witness: I swore something to that effect.

Mr. Feez: And Mr; Justice Windeyer said he absolutely disbelieved every word you said! Witness: I could not help that.

Mr. Feez: Did not the learned judge say so?

Witness: No,

Mr. Feez: Did he not say that you had committed perjury?

Witness: No.

Mr. Feez: You lost the case.

Witness: I consider we gained the case.

Mr. Feez: Did you not lose the case before Mr. Justice Windeyer and appeal to the Full Court?

Witness: Yes.

. His Honour: On what ground did you appeal to the Full Court? Was it on a point of law?

Witness: I do not remember, your honour?

His Honour: It is no use telling me that.

Mr. Feez: Do you mean to say that you do not recollect on what ground you appealed to the Full Court? Was it not on the ground that a declaration was not made before the marriage ceremony was performed?

Witness: It might have been.

Mr. Feez: It was a technical point?

Witness: I do not know.

Mr. Feez: You know what a technical point is. Your appeal was upheld on a technical point, and you cast off like a dog in the street the woman who had been living with you for about 13 years.

Witness: I provided for her.

Mr. Feez: Did not your uncle, the deceased (Mr. James Tyson), make you do it?

Witness: No.

Mr. Feez: Was not your wife living in Sydney in destitution until your uncle made you make her an allowance?

Witness: She was not destitute.

Mr. Feez: Do you remember the words of the learned Chief Justice when the appeal was upheld. He said, " I arrive at this conclusion, with deep regret inasmuch as the conduct of the petitioner all through is, in my opinion, reprehensible in the extreme." Mr. Justice Innes also said, " I need hardly say that I cordially concur in the hope that his honour the Chief Justice has expressed that the appellant, though successful in this suit, will not cast adrift in beggarly divorcement the faithful companion of well-nigh a score of years, united to him during that time, by what he believed to be the sacred bond of marriage, without at least making some adequate money provision for the remaining years of her life, and for their child. Indeed, I would fain cherish the hope that even at this late hour his instincts of manhood will prompt him to repair as far as in him lies the misery of this litigation by making her in legal relationship to him what she has so long morally been— his wife."

Mr. Feez: You married again, Mr. Tyson?

Witness: Not again; my first marriage was no marriage.

Mr. Feez: Oh, well, you can put it that way.

In further cross-examination by Mr. Feez. witness admitted that deceased was described in certain legal documents as of Felton, Cambooya. Witness looked on Felton as deceased's temporary residence, because he spoke of selling it. Deceased had better houses at Juanbung and Heyfield than he had at Felton Witness never heard the deceased use the word "home" except when referring to the place where his mother lived, which he had spoken of as " the old home." Witness would not admit that Felton was deceased's home: he resided more there than at any other place. Felton was the place that deceased set out from on his journeys, and the place to which he returned when he had completed them.

Brisbane Courier, **27 November 1900, P 7.**

Witness deposed that he was a nephew of the late James Tyson, and had managed Tupra station (New South Wales) for his uncle for twenty-eight years. For fourteen years prior to his leaving the station in 1892 he was also general manager of Heyfield (Tyson's Victorian station). After living for a time at Ivanbury, his uncle, in 1865 or 1866, acquired Heyfield, and built a house there, where he lived about two years, with occasional visits to Ivanbury. At the end of 1866 he built a new house at Ivanbury. Witness considered that both the Heyfield and Ivanbury houses had more conveniences than that at Felton. During witness's term as general manager, the house at Heyfield was only occupied by the deceased or himself on the occasions of their visits. Witness was general manager of all Tyson's stations, except those on the Darling Downs. None the less Tyson looked well after him - he kept his eye on everything. Witness was paid a salary for his work. The library Tyson brought to Ivanbury was left there, and the deceased's room was reserved for him. In the room was a tin box, where he kept his papers. Mr. Feez: Can't we see those papers?

Mr. Lilley: We haven't got them.

Mr. Feez: But they are in your power.

His Honour: They can only be got by commission.

Witness, further examined, said that Tyson invited him to live in Queensland, but he declined. He would not mind making money in Queensland, but he disliked the idea of living here. He knew, when he was well off. (Laughter). In 1888, Tyson told him he intended to exchange Felton with the Government for lands at Tinnenburra and Meteor Downs. Witness had never heard his uncle speak of Felton as his home. He had heard his uncle say to people who asked him where he lived, "Oh, I live everywhere." He spoke of himself as being "a bird of passage, always on the wing." Since 1861, when he sold two stations at Deniliquin, his uncle had disposed of no station or other properties in New South Wales. At a meeting of the relatives, held in connection with the administration of the estate, in Sydney, in December, 1898, a resolution was passed to the effect that Tyson's remains should be removed from Toowoomba to the family vault at Campbelltown.

By Mr. Feez: Tyson never said anything to him about being buried in the vault at Campbelltown.

Witness was then cross-examined at some length concerning a divorce case, in which he had been concerned some time ago, when he asked for a declaration of nullity of marriage against a woman who had been living with him as his wife for a number of years.

Further cross-examined, witness said that his uncle lived at Felton and died there, but he (witness) would not admit it was his home.

Francis Joseph West, Manager, Pilton
Telegraph, **15 November 1900, P 7.**

Witness stated that he knew deceased. Witness met him in the year 1870 on a part of Felton. Met him frequently afterwards. Had stayed at Felton as a visitor for about a week. Deceased spoke of Felton as the property he preferred living on out of all his properties. He also spoke of Heyfield in Gippsland, and said that although he had a very fine house there, he could not live there on account of the fogs. Witness had often asked deceased in a general way where his home was, and he always stated that he preferred living at Felton. Witness had met him when travelling, and if he was returning to Felton, he would say that " he was going home to Felton."

The witness, cross-examined by Mr. Lilley, said he had travelled with deceased from Sydney to Brisbane on one occasion. Deceased spoke of his properties and the number of stocks he owned. Witness asked him if a paragraph appearing in the newspapers about the number of stock he owned was correct. Deceased replied, "Not exactly; I own more sheep and less cattle than they give me credit for." He told witness that he would be willing to hand Felton over to the Government if they required it. Deceased took very little interest in local governing matters, except where he was personally concerned — that was with respect to taxation and roads through Felton.

By Mr. Feez: When witness met deceased on the steamer, he told witness that he was returning to Felton.

Charles Williams, Manager, Jondaryan
Brisbane Courier **14 November 1900, P 6.**

Witness said he remembered Tyson first going to Felton. Tyson used to speak of " visiting" his other properties, and of " returning" to Felton. Witness never asked him where his home was, but always assumed it was at Felton. He had visited there at Tyson's request, and was shown over the new house which Tyson had built after his own design. One large public room ran from front to back of the house, and the bedrooms opened off it. It was a peculiar structure, and Tyson was very proud of it.

By Mr. Lilley: He did not know if Tyson wanted to sell Felton, but he heard him say he would like to exchange it with the Government for freehold lands on Tinnenburra. Tyson had spoken of Felton as centrally situated to his other properties, and convenient for the despatch of his business.

Telegraph **14 November 1900, P 2.**

Stated that he did not know whether Mr. Tyson had any intimate relations in Queensland. Deceased had spoken of his properties in the other colonies. He had never told witness that he would sell Felton. He had spoken of exchanging it with the Government for freehold land on Tinnenburra. He had stated, that Felton was most convenient for him to do his business from, as he was in touch with the railway to the north and south, and in a central position to all his other properties. Witness thought for the last 16 years of his life deceased spent about seven months of the year at Felton. On an' occasion when witness had stayed at Felton there was a very good look there, witness had a very good dinner, there was a choice of spirits, and he thought it was very comfortable.

Henry Woolridge, Estate Agent, Deniliquin
Telegraph, **7 November 1900, P 2.**

Witness was examined on commission, stated in his evidence that he was an auctioneer, valuator, and estate agent at Deniliquin. He collected the rents of deceased's property. Witness had several conversations with the deceased and he always distinctly spoke of living at Felton. Witness Spoke to him about his property, and on one occasion asked him who he was going to leave it to. Deceased replied, " Leave it to the crows and let them fight for it.". He then explained that when hunting the kangaroos after being brought down by the dogs were left to the crows,' and so he intended to leave his property.

CHAPTER 6

Articles after the trial to determine domicile

Memoirs of a Stockman, **by Harry H Peck, P 81.**
Famous cattlemen and other noted suppliers

OF THE BREEDERS AND FATTENERS who regularly supplied Newmarket apart from the few already mentioned who generally drove their own stock, the most notable in the earlier days in the cattle line was James Tyson and, later, taking Tyson's place as the Cattle King of Australia, Sir Sidney Kidman. I knew these outstanding cattlemen well for many years.

... (Circumstances of the Tyson family coming to Australia has been deleted as proven wrong.)

Nothing much is known of his education, but that he received, or adapted himself to a good one is certain. For though in his later years his manager, Buchanan, at Felton, his home station on the Darling Downs, wrote nearly all his letters, his own letters, in a firm flowing hand, bore the stamp of such a man. Tyson stood six feet two inches in his socks and was broad shouldered and deep chested in proportion. It took a good horse to carry him well. When I first met him, in the '70s, during one of his periodical visits to his famous fattening property, Heyfield in Gippsland, "The Old Gentleman" was handsome, straight as a rush, with a long flowing iron-grey beard, and would have been a man of mark in any crowd. However, he loathed personal publicity and busybodies who tried to draw him out, and consequently travelled incognito as "Mr. Smith" or "Mr. Shiels". But it was not of much use, as once met or seen few were likely to forget him, considering his reputation and his unusual personality. He was a life abstainer from alcohol, of which he had a dread, and a non-smoker, although he usually carried a plug to use sometimes for snuff, or give a traveller a pipe-full. Because he never married, or was ever known to have any love affair, folks sometimes called him "a woman-hater", but in this they were mistaken for, though generally shy, he had a genuine regard and friendship for the few managers' wives or housekeepers on his stations, who had a chance of looking after his meagre wants in his later years.

His philosophy as regards women in general, however, was unique. He thought that at the time of full development all should be medically examined and those who did not come up to standard in development and constitution, as likely to rear well, strong and healthy children, should be sterilised, and the race thereby be improved. His argument, "we cull our breeders to improve our stock, why not improve the human race on the same lines", was logical, and that system he thought civilisation might some day have to apply.

On one of his visits to Heyfield he went across to a special store cattle sale at Maffra, which in those days was the principal store cattle market of Victoria. He returned to the city the evening of the sale by train with his Melbourne agents who, with Mr. Tyson's consent, invited several other stock salesmen to share their compartment. They were only too ready to accept the invitation. Among the number was Tom Mates who, before turning stock agent, had been first a drover and then a fat cattle dealer, largely on the head-waters of the Murray and Murrumbidgee. Tyson in the early years of Bendigo supplied the diggers there with meat and, after cleaning Riverina out of fat stock, had made a visit to the Upper Murrumbidgee in search of further supplies, and taken with him a dray heavily laden with stores for rations. He had never been in that district since, and told Tom his clearest recollection of it was an exceptionally steep and long hill on the track. He said that he had the greatest difficulty in getting his heavily loaded dray to the top, having to heave his shoulder to the wheel of the dray, first on the one side and then on the other to help the horses, as it slowly criss-crossed and crawled up the hill. When he got at last to the top, he was nearly exhausted, but was given refreshments and welcome hospitality in a little wayside shanty by one of the finest women he had ever set eyes on.

Then the old gentleman described her, and his description was one of those rare occasions on which he came out of his shell and fairly let himself go. She was his ideal woman, tall, clear eyed, rosy cheeked, the embodiment of blooming health and constitution. The agents, who had met him for the first time, listened with surprise as they had always thought him to be "a woman hater", but Tom Mates, irrepressible as usual, and who would have not been abashed in the presence of an Emperor, said: "Well, Mr. Tyson, I have always heard that you never tasted alcohol, smoked tobacco or kissed a woman in your life, but what about that lass at the top of the hill? I guess you relaxed your rule in the third degree that time". Tyson made no reply and, treating the question with contempt, went right back into his shell.

His elder brother William, with James Tyson, pioneered the Lower Lachlan, when practically in their boyhood, taking up those huge areas Tupra and Juanbung, having followed that river down from its headwaters until they came to unoccupied country. An old friend of their father, admiring their pluck, gave them a few head of cattle to start stocking up, and it fell to the lot of James to go back to their home district for them. He had one shilling to finance the trip, and when he came to the Punt at Forbes the river was in flood. The punt man offered to take him across for the shilling when he found that was the extent of the lad's finances, but James decided to keep his shilling and swim for it. He did, but nearly lost his life in the crossing. He got the cattle and landed them safely down at Juanbung, a good 500-mile journey from there and back, and the shilling still in his pocket. That trip was characteristic of James Tyson as a young man, when he denied himself many a necessity in order to save money, and no doubt was the origin of his repute as a "possum eater", of which he was not ashamed, for many another good and hardy pioneer had eaten 'possum and, when properly grilled, relished it, too.

For many years the progress of the Tyson brothers was slow. Cattle were low in price and almost unsaleable, so they made cheese by very crude methods, and in a very unsuitable climate, and twice a year took a dray-load over the hundreds of miles of rough tracks to Sydney, returning with the proceeds in stores. Meanwhile, their herd gradually but surely increased in numbers, and then in the '50's their big chance me. Gold was found, and Bendigo, only about 200 miles by a good stock route from their runs, soon had a population of many thousands, for which meat was the staple diet, and after a good season their cattle were all fat. Instead of selling to dealers or butchers they built an abattoir outside Bendigo, which William ran, while James did the droving. They sold direct to the diggers, first come first served, at tail prices, and made a fortune out of their own stock. When their own fats were used up, James bought far and wide both sheep and cattle and, after cleaning out all who would sell in Riverina and central New South Wales, went right up into southern Queensland, which had only recently been stocked, for supplies. On these droving trips from the north to Bendigo, though by that time practically coining money, James further enhanced his reputation for economy, as often his only helpers were black boys.

When Bendigo was played out, after Henry Ricketson and other dealers came in, the Tyson brothers cleaned up and cleared out. James' share of their Bendigo operations was £200,000. This set him up financially and was the basis of the huge fortune he made in stations. He bought William out of Tupra and Juanbung, and then began ending his Bendigo profits to other squatters, often without any legal security. Station after station thus fell into his hands, as once Tyson lent money on a property luck seemed to desert it and bad seasons set in until he took over, when the hoodoo would lift. For instance, the owner of a big cattle station in western Queensland who had, in Tyson's latter years, borrowed from him on his run and herd, had such a prolonged drought that in disgust he wrote Tyson several times to come out, take over and give him a clearance; but, failing to do so, the owner at last cleared out in desperation. Within a week the long drought broke and Tyson got the run and the best Hereford herd in western Queensland.

On one occasion, when making a search through old papers left on me of his stations, the manager and I came across a number of cheque butts of the '60's and '70's, and were surprised at the number of loans they represented to well-known station holders. Amounts ran up to tens of thousands each, and at interest generally around 7 and 8 per cent. Apparently, many of such loans were without security, as memos m the butts simply read, "Lent to Mr. So-and-So at so much per cent., to be repaid by proceeds of fat stock or wool", as the case night be.

Like most bushmen, Tyson was a great tea drinker. At Felton on the Darling Downs, which he made his home for the last 20 years of his life, and where he considered the climate "the best in Australia", he had his tea served in a billy on the sideboard. when an old friend inquired the reason, he said, "My housekeeper used to serve my tea in a tea-pot covered with a night-cap (he referred to a cosy), and several times I burnt my mouth from the over-hot tea. I asked her to cut out the night-cap but, as she persisted, I eventually threw tea-pot and night-cap out of the window and have had the billy ever since, and tea as it should be". The collectors for a new church on the Downs approached James Tyson for a donation to head the list. He asked the amount of the estimated cost, got his cheque book and, to their surprise, gave a cheque for the lot. When the church was nearly finished it was found the architect had omitted to allow for the lightning rod, and the church committee went back and asked Mr. Tyson would he like to complete his gift by paying for the lightning rod. They got a curt refusal in these words: "No, if God thinks proper to strike the church with lightning, I am not going to attempt to interfere".

Tyson began as a cattle-man, and his main interest was always cattle. Until his death in 1898 he was recognised as the "Cattle King" of Australia, and the man who made the most money out of cattle. He made much money out of sheep but, like Sidney Kidman, who succeeded him as the "Cattle King", and who extended his cattle interests to much larger dimensions than Tyson's, sheep were only a sideline with him. On Tinnenburra, a huge holding and the largest of his stations, which extended from the Warrego on the east to the Paroo on the west. Musterings, owing to seasonal conditions and the great area of the place, were sometimes very incomplete. The result was that often old bullocks up to 10 to 15 years old, which had perhaps never been mustered since they were branded as calves and had become very wild and "real old pikers", were sent down to Heyfield to fatten. The Heyfield managers regularly protested but in vain, for the "Old Gentleman" always replied, "Heyfield can fatten anything". He was about right, for in those days T. K. Bennett and John Woolcock fought each other keenly for the heaviest prime bullocks at Newmarket. The big weighty extra prime TY1 brand Tinnenburra-bred bullocks from Heyfield, loading only eight to the big truck, often averaging 1000 to 1200 lb., and constantly topping the market, were bullocks to conjure with, and a great pleasure and pride for any auctioneer to sell.

A few years before his death, following a drought on Tinnenburra, and the raising of the Victorian Border Tax on cattle up to 30/- per head, Tyson gave his then manager at Heyfield, Mr. Charley W. K. Whittakers, permission to buy some hundreds of Victorian-bred store bullocks. As the raising of the Stock Tax (which fortunately Federation wiped out) made it impossible to buy New South Wales or Queensland bred stores, Mr. Whittakers' choice became very restricted, so to stock he had to include in his buying some of dairy-bred blood. On Tyson's first visit to Heyfield after the purchase, Mr. Whittakers took him through the various paddocks in which the Victorian-breds were running and, after seeing them all, he turned to the manager and said: Well, Mister, I think them a lot of 3 B's". As known to cattle-men, B's means browns, brindles and bastards. Tyson, however, was never known to swear, and in fact was most courteous in his conversation, though he could be very cynical at times. He had a peculiar habit, too, more particularly in later years, when deafness and an inability to catch or remember even familiar names quickly, affected him, of addressing almost everyone as just plain "Mister".

On Tinnenburra Tyson had more artesian bores than any other run in Australia, and some of them of great depth, with bore drains running out, and watering otherwise waterless country, for 20 miles and more. At the time of Tyson's death, Tinnenburra was the most expensively and best artificially watered property in Queensland. whatever improvements he made on any of his stations were always substantial and made to last for generations, the yards, fences and dams on Heyfield being an object lesson to Gippslanders. He never begrudged spending money on good improvements where necessary. Tyson was very proud of his bores on Tinnenburra, and of being one of the first to realise what the tapping of the great artesian basin meant to pastoralists; in fact, he had a great appreciation for engineering in every form, and often regretted that as a young man he had not had the opportunity of a technical education. He thought so much of the value of such a training, more especially as applied to mechanics, that when an old friend and trusted agent visited him at Felton only few weeks before his sudden death, he said that he had so far made no will, but that after due and ample provision for his relations, he tended to will about half of his huge estate to found and endow a technical college for the education in mechanics and engineering of boys intending to go on the land. This was to be known as the "Tyson Institute". Alternatively, he said he would leave the bequest to the Governments of Queensland, New South Wales and Victoria, to be funded, and on the anniversary of his birthday in each year the interest would be divided between the charities most in need at the time in the respective States, under the title of the "Tyson Bequest". He asked his friend for his opinion as to which was the better plan. The friend replied that he considered both were admirable, but that as it was not his money, he did not wish to give an opinion, and Mr. Tyson should decide the matter for himself.

However, he realised how ill the old gentleman then was, although Tyson himself thought he only had a heavy cold, and strongly advised him to see a doctor without delay.

Tyson agreed to do so and, after his friend's departure, told the manager of Felton, Mr. Buchanan, to drive him about 20 miles to consult the doctor at Pittsworth. When about half way he asked Mr. Buchanan, "Where are we going?" and on being told, "To Pittsworth to see the doctor", he replied, "Pull up and drive back home; I have never consulted a doctor in my life and am not going to do so now". Although very weak he never laid up, and a few days later died during the night in his sleep, but without making his will. About 50 relations divided his estate under "next-of-kin" conditions. The estate realised £2,500,000, but he died in 1898, when cattle were cheap and wool was low in price, and consequently his stations were difficult to sell. But before the closing and reconstruction of the banks in '93, James Tyson must have been worth nearly £4,000,000. In his later years the Queensland Government made him a member of the Nominee Legislative Council or Upper House, but he seldom attended, and was never known to make a speech.

Two anecdotes are related by an old Queensland friend:— Following a good season, a boom in cattle and wool and a strong demand for stock, a friend meeting him in Brisbane said: "I don't hear of you running around like so many just now on the buy, and you must have lots of feed to spare". Tyson replied: "When others run, I walk". On an exceptionally cold winter's night in Sydney for that climate, when a cold westerly was blowing off the snow of the Blue Mountains, and during a really bad season and low prices, a bunch of outback squatters, including Tyson, was sitting around a fire at a well-known meeting place. Generally bemoaning the hard times, but with Tyson, as usual, only listening, one of them said: "Well, I think the country is going to 'the dogs', but we haven't had your opinion, Tyson; what do you say?" Tyson replied: "I have heard that lament many times in my lifetime, but I haven't yet heard them barking".

Sydney Mail, Wednesday, 18 December 1907, P 1575.

MR. JAMES GORMLY'S RECOLLECTIONS

Mr. James Gormly, M.L.C., writes:—
"I have been much interested in reading the reminiscences of Mr. Edmond Morey that have appeared in the 'Sydney Mail,' as I know many of the old hands referred to in the articles. My father settled on the Murrumbidgee River in 1844, and our home was situated on the overland track to the province of South Australia. From 1844 to the time of the discovery of gold, numerous herds of bullocks were taken overland from New South Wales to South Australia, and were purchased there, and made use of to haul copper from Kapundah and Burra mines to Port Adelaide.

"Darchey and his wife frequently called at our home when going to his station at Oxley, on the Lower Lachlan, and I heard him and many others refer to the treachery of the blacks in that part of the country.

"Although I was only a boy in those days, I was frequently on the road with stock, and I can confirm the statement of the writer of the articles that the blacks were in the habit of holding spears between their toes as they approached an unsuspecting white man, and when near him raise their foot to their hand, and hurl the weapon at the man. That system of treachery was quite frequent on the Lower Murrumbidgee and Murray, and only that I was on the alert, I would on more than one occasion have been likely to suffer from this system of cunning.

"I first became acquainted with James Tyson, his mother, and other members of that family in 1843. They then resided on their farm, which was close to where Major Mitchell erected Nepean Towers. I well remember when the Major first began to clear the scrub from his land.

When the Tysons first took stock to the Lower Lachlan, they had only a small herd, and their cattle were chiefly poddies (milkers' calves). When Jem[12] first took a small mob of fats on the road to the Sydney, market he put them in my father's stockyard one night. He had no assistance to drive the stock except a dog, and he had only one horse, which carried himself and his swag.

"When gold-mining in Bendigo in 1852, my brother and myself had our tent pitched close to the yard where Jem Tyson and his three brothers were slaughtering stock and selling quarters of beef to the retail butchers. I have known those men to work from daylight in the morning to well into the next night.

"In article No. VI., in the concluding paragraph, the writer is wrong in stating that the channel which connects the Murray River with Lake Victoria was named the Rufus in consequence of the waters having been discoloured with the blood of about 40 blacks, who were shot while swimming over the stream. Captain Charles Sturt, in Vol. II. of his account of the expedition, he led down the Murrumbidgee and Murray in 1829-30, writes on page 128:— 'I called the little tributary the Rufus in honour of my friend M'Leay's red head.' George M'Leay was afterwards the owner of Toganmain station, on the Murrumbidgee, and I have heard him relate many incidents of that memorable and successful expedition from Sydney to the mouth of the Murray."

---

[12] It is reported James' mother called him Jem but she was illiterate so she could not write 'Jem' rather it was her accent.

Sydney Mail, 13 Aug 1913, P 8.

Random Recollections
By (the late) J.E. Richter
A MEMORY OF JAMES TYSON

As a variation to his gold-mining experience, the writer at a later period was engaged as a cattle drover to James Tyson, who was then fast becoming the wealthiest man in Australia. It was droving a lot of fat cattle from one of his stations near Walget to be delivered to one of his agents at Bendigo. He accompanied us a part way at the commencement of the journey and took quite a monopoly of the conversation of the writer for no reason that was at first apparent. Whatever was the subject of the conversation at the start, it was always veered round onto geology – a subject I had given much study to and was fairly well-versed in. In later years Tyson made the study of geology a hobby. At first, he had studied it to ascertain if it would assist him to find artesian water to increase the value of his many pastoral properties. I quite well remember him pointedly asking me why the primary rocks were unstratified. That question, I replied, had never been satisfactorily answered: different men had different theories on the subject. It set him thinking for several days. Then he asked me if the granites had any influence on the production of artesian water. "None at all," I answered, "except so far that they may in certain cases act as an underground dam and thus form an artesian basin." His next question was, "Is that apparent on the surface of the country?" My answer was, "Yes, in certain cases. If the tertiary strata have a dip from a higher country down a broad valley that may be 100 to 200 miles wide and a granite formation shows outcrops on any low hills across the line of that valley, thus forming an underground dam, then above that line will be found artesian water, provided the high country is situated in a region with an annual rainfall sufficient to make the said artesian water. If the high country was situated where the rainfall was but five inches per year, this quantity would be evaporated or soaked up in the surface soil and therefore would never reach low enough to form the artesian water, as in the deserts." And then, like a hen standing on one leg, Tyson was set thinking again.

AN INTERESTING PERSONALITY

James Tyson dying at the age of 78 (sic) years and leaving an estate valued at three and a half millions was not an ordinary man. He was 6 feet 4 inches in height and well-proportioned with intellect above the average, but handicapped by the want of higher education. The writer first became acquainted with him about 1848 and was by chance thrown into his company many times in the ensuing 40 years. He was the most interesting personality that ever came into my life. He did not smoke tobacco which he designed a vile allurement. Nature never intended a man's mouth to be made a chimney of, is what he had to say. Nor did he indulge in spirituous liquors to steal away the little brains apportioned to mankind. He did not swear, or bet, or gamble. He never married and perhaps for that reason never went to church. He never patronised the theatre or the racecourse and preferred to sleep out in the open air, often with a saddle for a pillow, not to be confined to a house with fine bedclothes and linen. He talked but little, "the person that talks much cannot talk all sense," was one of his proverbs. He treated the aborigines with a kindly regard and claimed that they were more sinned against than sinning. The deserving white man or woman, fallen a prey to ill-health or misfortune, claimed his attention; but woe to the crafty white man who tried to get the better of him by fraudulent means or otherwise. Of high integrity himself, he brooked no mean action in any other person. Of simple tastes and habits, he eschewed the ranks of aristocratic society. He was more than once invited to

Government House by Lord and Lady Carrington but did not accept the invitation. "Their ways are not my ways and I should only be like a fish out of water amongst them," is what he was heard to say in reference thereto. "Besides I cannot see that such a visit would be of any benefit to me in any way and they would expect me to admire their bric-a-brac, fine feathers and geegaws. It is not for me - no, it is not for me." Did he ever think of marrying? Yes, he once took a strong notion of a girl. The daughter of one of his father's neighbours, but he was a busy man then - too busy to ask her - and when he did resolve to ask her to become his wife, he found that the blacksmith had already married her. Make money? Oh yes, he loved to make money – it was the elixir of his life. Make a will? No, he would not make a will. It would not be fair, as he had not seen and knew nothing about his relatives.

TYSON ANECDOTES

Many peculiar anecdotes have been told of Tyson that have already found their way into print, but here two new ones. Once on visiting Appin, the scene of his boyhood days, he met a young unmarried woman who he had not seen for several years. They were shaking hands and he was raising her hand to his lips to kiss it when he was seized with the convulsion of a violent sneeze during which their hands were loosened apart. "The fates are against me, you see. Poor fellow me, poor fellow me!" he said. The latter words were a favourite expression of his used in imitation of an aboriginal wail. On another occasion, having been absent for a few months he approached one of his own stations disguised as a swagman. Going up to the storekeeper, who was a stranger to him having been appointed by the manager during his absence, Tyson accepted a double quantity of rations tended him to take to the men's hut, where cooking and sleeping accommodation was found. The rations given free to all swagmen were a pannikin of flour, a pound of meat and some tea and sugar, according to Tyson's standard orders. After breakfast next morning, Tyson, dressed in a different style, appeared with the manager before the storekeeper to demand why he had exceeded instructions in doling out a double quantity of rations to a swagman the evening before. "Because it appeared to me that you really needed it, being such a big man," replied the storekeeper. "Well," said Tyson, "have a care and don't do it again," His transactions in stock sometimes assumed very large dimensions. On one occasion he gave a cheque for £72,000 for cattle purchased. When the writer knew him, he had not yet left his parents at Appin, although over 21 years of age. Whilst with them it was his duty to drive two or three horses in a dray to Sydney. On the way he sometimes called at my father's flourmill at Liverpool, leaving several bags of wheat to be gristed into flour ready for him to take home again on his return. He first commenced his career by buying and selling a few cows about Appin, leasing grass from his neighbours for their sustenance whilst in his hands.

The quotation below is interesting for a number of reasons. Isabella Dwyer was a niece and possibly the eldest in the family at the time of publication, which is the main item of interest. His attitude to females strongly enhanced views expressed by others but I would add that he is reported as saying he promoted marriage amongst farm hands to keep them in check. The contention that James did not like managers having wives with them is true when I think about it and it is probably because he often stayed in the house. She further verified James had some formal schooling (See Chapter 11). She showed she gave air to the rumour that James was a poet (see Chapter 10). She was surprisingly off the mark with where James died as it has been proven to have been at Felton. Maybe a small slip-up in her memory which would have been picked up by an Australian journalist who would have known Tupra was a long, long way from Brisbane!

Northants Evening Telegraph, Saturday, 29 June 1901, P 8.

TYSON'S MILLIONS
HIS NEXT-OF-KIN GOES TO LONDON AND TELLS THE TRUE STORY OF THE LONELY SQUATTER.

The case of the eccentric millionaire squatter, James Tyson, who died intestate in Australia, and about whose five million so many strange stories are told, is recalled by the visit to England of Mrs. Dwyer, one of the eight persons amongst whom the Australian courts have decided the money shall be equally allocated, for further division amongst their families. Mrs. Dwyer, besides being a niece of Tyson, the son of the Cumberland emigrant, has a closer connection with England; for her daughter married the son of Mr. Charles Morton, manager of the London Palace Theatre. Mrs. Dwyer and her two daughters are at present on a visit to Mr. & Mrs. Morton; and she told a Press representative the life history of the strange old cattle man.

Mrs. Dwyer is Australian born and bred and she knew Tyson well. She saw him regularly on his bi-annual visits to the house outside Sydney where she lived. Her first recollection of him were woven with stories of his eccentricity, "Perhaps what seemed the strangest thing about him," she said, "was his extraordinary aversion to women's society. He was fond of his own people but he would not have anything to do with anyone else. He allowed none of his managers at the different cattle stations to have their wives with them[13]. If a woman got into a railway carriage in which he was travelling, he went into another compartment.

"He started making his money in the Bendigo fields, about '52

---

[13] In the Domicile trial evidence was given that he liked workers being married apparently not so with managers

when the rush was on. He had worked for some years as an ordinary cattleman up country; and had saved enough to start with a few beasts. When the gold fever came, he drove them over to Bendigo and sold them at a profit that enabled him to buy up a fresh lot. He never stopped making money after that.

"At one time he was worth ten million; but he was hard hit by some big Australian bank smashes years ago. It was one bank that failed he had a half a million of money standing to his credit account.

In his prosperous days he would give big sums in a most eccentric ways, He would shake hands and leave a ten pound note behind—always notes, never cheques so few knew what he really gave to others. When the Soudan War was on, he gave £2,000 towards the fund for sending our Australian soldiers and offered £2,000 a year until the war was over.

"When he did get the money, he did everything in a big way. His stations in Queensland were never smaller than 40,000 acres; and he boasted that he would not own one worth under £100,000. Since his death one was sold for over that amount. Another, called, Glenormiston had 2,500 square miles of land in it.

"They called him 'Hungry Tyson,'" Mrs. Dwyer said, "because he deprived himself of the luxuries living. When he had a pound, he did not spend it; but waited until he could put another to it. He made himself a regular weekly allowance, and he never went beyond it; just as if he weren't worth a penny. But there was never a better son or a better brother. It was only to women that he showed any strange aversion. The family never knew of any reason for it. He had never had any love story as a boy at school, he had the same dislike to feminine society.

"A favourite dodge of his was to go dressed as a workman to one of his stations where there was a new manager. He would work, sleep, and eat with the other men, and see how everything was going.

If things were all right, he would pass on without disclosing his identity; if he didn't like anything, he paid the manager his wages next morning, and packed him off.

People called him queer things for that way he went on. He was once called 'Tyson the poet'; and in his pocket-book after his death was really found a poem by him[14], beginning—

Let the wealthy have their customs
    And keep the wealth they own:
But with claiming all their pleasures
    Let other men alone.

"That was what he wanted. He made his home (when he was there) at a lonely and secluded station called Tupra. He had bought the land; and because it was dry had turned a river across it! The Government stopped that sort of thing by special Act afterwards. But Tupra was not the richest of his stations. For one called Tinnenburra he refused a million of money.

"He died as he lived. He always stayed at a club when he was in Brisbane; and there he was taken ill. His solicitor advised him to see a doctor, but he refused and packed his traps and went to Tupra. Even there he would not take to his bed. He was walking about on the Saturday night. On the Sunday morning they found him dead[15].

"Before his death he said to his manager, 'There'll be some fun after I am dead'—referring to the contention there would be over his will. He knew he had made no disposition of his property. He drew up wills for all his brothers; but never made one for himself."

---

[14] See chapter 10, the poem was by somebody else

[15] James died at Felton, verified from many sources

Queensland Country Life, Thursday, 8 March 1951, P 9.

Pastoralists I Have Known
James Tyson

My first meeting with James Tyson was when I was a passenger on Cobb's coach bound on a visit to a relatively close to a holding of Tyson's.

My first impression of him was that of a forbidding man of large proportions with a thick beard but well dressed and wearing what I found out later was the custom with men who travelled the outback tracks—a "wide-a-wake" hat.

It was some years later when I again saw him. That was when I joined up with the Kennard Marwedel outfit in Pittsworth and his home was close by at Felton. He was often in the township, mainly to send telegrams and to yarn with the proprietor of the Buaraba Hotel, a Mr. William Bowden, who was his first manager at Felton and came with him from the South.

The hotel was one of the foundation buildings of what was then known as the Buaraba township. It still stands and must be one of the oldest licensed premises In the State.

When later I went to the Central district, I saw Tyson again. He was visiting Meteor Downs, near Springsure, which he owned. Although I never actually came in contact with him in business in Plttsworth or in the Springsure district, the businesses I was connected with did and I formed an entirely different opinion from that held by a lot of people who were never in contact with him. The general idea and the reputation built up have been that he was a miserly old curmudgeon, not by any means cultured. This was entirely wrong. Because he would not put up at bush pubs or set up drinks for the crowd in the clubs or hotels he patronised in the cities or larger towns. He was not popular with the majority of his fellow pastoralists, who liked the fast life of that day. On the contrary, his intimates, while quite aware of his exactness in money matters and practical economy in the management of his properties, would never subscribe to his being termed a miser.

On matters of public interest, he was well informed and was widely read on solid literature. When he was raised to the Upper House, it was because the Government of the day wanted his advice on pastoral matters.

CONTROLLED MANY LARGE RUNS

Tyson was always well dressed. In the cities he always put up at the leading clubs or first-class hotels and travelled first class on the railways and saloon on the interstate boats. In Brisbane his headquarters were the Queensland Club and in Toowoomba the Royal Hotel, which at that time was the headquarters of leading pastoralists on the Darling Downs and further west.

In his business he was peculiar. His stock affairs in this State he left in the hands of the late Harry Bracker, manager of the stock department of Moreheads, and in the later years of his life he shared some of it with the late Alex McPhie. These gentlemen gave him the reputation of being an easy man to sell stock for. In his operations with different business firms, he insisted on quality and service instead of cheap methods.

From my own knowledge of his pastoral properties, he was one of the most up-to-date in his improvements. The wool sheds on Tinnenburra and Meteor Downs, and the water distribution scheme on Mount Russell, were monuments to this trait in his character. His gateposts and fencing were of the most substantial materials procurable.

In the early 70's he turned his attention towards Queensland and records in my possession show that a large number of runs were registered in his name. On the Eastern Downs he held Buaraba, which comprised practically all the area east from Pittsworth to South brook and north of that line to the boundaries of the old Westbrook run. "When the township now known as Pittsworth was formed by the establishment of a hotel (the latter now standing and still licensed) it was known as the Buaraba township and the name Pittsworth only came when the railway line was first mooted, it being derived from the maiden name of the wife of the Hon. Macdonald Patterson, a Minister who was the main advocate in Parliament for the railway line and was also Tyson's solicitor. The lady's name was Pitts but the addition of "worth" I never heard explained.

On the Western Darling Downs, he held Arubial, Bentland, Myall Grove and Miamba. On the lower Warrego over 40 blocks stood in his name, aggregating just on 3500 square miles by my records. This afterwards was consolidated into what for about the last 70 years has been known as Tinnenburra. It stretched from the Cunnamulla Road to Eulo along the southern boundary of Bowra to the western end of that run in a north westerly direction for about 10 miles and then southward to the N.S. Wales border and then east to the Warrego River and then north on the western bank of that watercourse to Cunnamulla,

Names of many of the blocks are still maintained on grazing farms taken up at the first resumption, such as Moonjaree, Tuen Plains, Mowellen, The Gum Holes and others.

TYSON SAVED THE BANK

In the Central West Tyson also had a large holding. On the lower Thompson he held the blocks Warbreccan, Mutti, Corella, Jundah, Garvoe, Holbern Hill, and Warungle, all of which were afterwards consolidated into the holding we now know as Warbreccan.

This country, with the exception of Tinnenburra, was not stocked by Tyson. At the time he was a large money lender and the prevailing idea was that they were mortgaged to him. Then and for many years afterwards it was the custom for mortgagees to have pastoral leases transferred to them, as it was held that the mortgaging of leases held from the Crown were not valid. All the properties held by Tyson in this State, with the exception of Tinnenburra, fell to him as the mortgagors failed to meet their commitments.

One of Tyson's finest acts in this State—and according to his intimates there were many—was the saving' of the Queensland National Bank after its closure; the result of the 1893 bank smash.

My information is that when discussing with his friend Drury, who was the general manager of the bank, how it could be saved, Drury said-this could be done only by being able to secure at least £100,000 worth of Treasury notes which the Queensland Government was then issuing, but this would take that many sovereigns. Tyson replied that he would see what he could do. Now as far as I am concerned, this latter is only hearsay, but here are the facts as to how Tyson helped: The Australian Mortgage Land and Finance Company Ltd., as it was known then, was approached by Tyson to get its London office to forward 100,000 sovereigns at the first opportunity and these are the facts as I had them from the late W. P. Devereux who was, for many years, the manager of the company in Brisbane and who at the time was in the company's London office:—

When the cable was received in London the British India steamer for Brisbane was scheduled to leave Gravesend on that night's tide which meant that the gold had to be got from the Bank of England and taken by van to the point of despatch. As the journey would be through an unsavoury locality, several of the senior officers of the company demurred at doing the job and Devereux volunteered. When the gold reached Brisbane, the bank was able to obtain the Treasury notes. The situation was saved and the bank has never looked back.

Another service he did the nation was when the N.S. Wales Government sent the contingent to the Sudan War. There was an anonymous contribution of £2,000. Tyson turned out to be the donor.

CHAPTER 7

RUMOUR: James Tyson never left Australia

In late 1896 and early 1897 most Australian city and country newspapers had a snippet saying James Tyson was going to England for Queen Victoria's diamond jubilee celebrations; some even said he was going to be knighted!

Brisbane Courier, **Thursday, 4 March 1897, P 4.**

The arrangements for the Premier's visit to England have now been completed. Sir Hugh Nelson and his family will go by the Norddeutscher Lloyd steamer Barbarossa, which leaves Sydney on 27th instant. The Hon. T. J. Byrnes will also go on the same boat. Mr. H. S. Dutton, secretary to the Chief Secretary, will accompany Sir Hugh. It is understood that the Hon. James Tyson, M.L.C., will also visit England on the occasion of the Diamond Jubilee celebrations.

Sydney Morning Herald, **Thursday, 4 Mar 1897, P 4.**

Mr. James Tyson has decided to visit England in company with the Premier of Queensland. He has given £100 towards the sending of the Queensland rifle team to England, and states that if it be successful he will give another £100.

The Age, Wednesday, 2 June 1897, P 5.

It seems to be taken as an assured fact by the London press that Mr. Wilfrid Laurier, the Canadian Premier, who originated the new Canadian tariff granting preferential treatment to British imports, will figure on the list of Jubilee honors. "Anglo-Australian," in the "European Mail," gives the following item: — "Mr. James Tyson, the well-known Australian squatter millionaire, is coining to London for the Diamond Jubilee celebrations. Mr. Tyson is generally admitted to be the richest man in Australia. He is a member of the Legislative Council in Queensland, and in an unostentatious manner has done some valuable service in the financial way for that colony. It is not unlikely that Mr. Tyson will be knighted this year."

The following appeared in print. All the details must have been finalised and the fare paid. James was almost on his way in a bed in the luxury he professed to abhor. He definitely would be out of place sleeping on the deck of the Barbarossa.

Warwick Examiner, Saturday, 6 March 1897, P 4.

The Hon. J. Tyson, M.L.C., has decided to visit England, and will accompany the Premier on his approaching trip. The arrangements for the Premier's visit to England have now been completed. Sir Hugh Nelson and all his family will go by the Norddeutscher Lloyd steamer Barbarossa, which leaves Sydney on 27th inst. The Hon. T. J. Byrnes add Hon. James Tyson will also go on the same boat. Mr. H. S. Dutton. secretary to the Chief Secretary, will accompany Sir Hugh. Mr. Tyson has donated £100 towards the fund for sending a Queensland rifle team to England, and he states that, if the necessary amount is raised, he is prepared to give another £100. It has been finally decided to send the team.

They were even getting excited in Britain. In the Daily Mail, circulating in Hull, East Yorkshire and Lincolnshire, had the following prominently displayed on the front page:

The Daily Mail, **Monday, 26 April 1897, P 1.**

AUSTRALIA IS SENDING US FOR THE DIAMOND JUBILEE HER IDEAL MAN, MR J. TYSON, THE FAMOUS SQUATTER MILLIONAIRE.

The news even went to Scotland:

Dundee Evening Telegraph, **Tuesday, 27 April 1897, P 2.**

One of the most interesting announcements connected with colonial visitors to the jubilee is that James Tyson, the multi-millionaire and erstwhile champion miser of the Southern Hemisphere, is coming.

Mr. Tyson is a man with an eagle eye and a single mind to making money. He made his first great haul shortly after the great rush to the Bendigo goldfield. Thousands of gold seekers were encamped on the golden flat, and a meat famine was at its highest when Tyson shrewdly arrived on the scene with a herd of cattle and a numerous flock of sheep.

He made a profit of something like 10,000 per cent on that little speculation, and he went on investing in sheep farming, railway contracting, acquiring land in the vicinity of rising towns, &c., until now his accumulated wealth is variously estimated at from five to seven million.

But, then, this appeared:

***Telegraph*, Brisbane, Saturday, 13 March 1897, P 6.**

The Hon. James Tyson, who had bespoken a berth in the great German steamer Barbarossa, has, it is said, relinquished the berth and abandoned his intention to visit England this year.

Mr McIvor said in his testimony James dearly wanted to see the Queen but the crowds and the 'swells" worried him. The concern must have got to him so much that he relinquished the opportunity to meet the Queen. Other testimonies reinforced the view that James was shy and retiring and he felt uncomfortable around people, particularly among females.

So, the Premier of Queensland departed with his family, as shown in The Daily Telegraph, without James Tyson or his aliases, Mr Smith, Mr Walker or Mr Stokes listed.

***Daily Telegraph*, Saturday, 27 March 1897, P 11.**

The following passengers have been booked by the Norddeutscher Lloyd line mail steamer Barbarossa, leaving the company's wharf, Circular Quay, at noon to-day for Southampton and Bremen via ports:—

From Sydney for Southampton, London, Antwerp, and Bremen: **Sir Hugh and Lady Nelson, Misses Nelson (three), Mr. Duncan Nelson, and valet,** Hon. T. J. Byrnes, Mr. and Mrs. H. S. Button, three children, and nurse, Mr. and Mrs. W. J. Overell, Mr. H. C. A. Hyde, Mrs. J. K. Meredith and two children, Mrs. M. A. Havard, Miss K. M. Havard, Master Havard, Mrs. D. F. Brown, Mr. and Mrs, J. D. Ridley, Mrs. F. Hicklin and two children, Mr. J. G. Broom, Mr. C. Baldwin, Mr. and Mrs. J. S. M'Donald, Mr. J. Hopson, Miss M. Pagan, Mrs. W. Courthope, Mme. Leverrier, Mdlle Yvonne Leverrier, Mme. Louise Leverricr, Miss Jeanne Leverrier Masters Harry and Maurice Leverrier, Mr. and Mrs. F. E. Joseph, Miss and Master Joseph, Mr. and Mrs. P.

Pfleiderer, Mr. K. and Mr. W. Pfleiderer, Mr. W. A. Benn, Rev. Gregorius de Groote, Rev. M'Carthy, Mr. T. T. Dixon, Mr. and Mrs. Robert Chadwick, Mr. Herbert Chadwick, Mr. W. F. Fox and servant, Mr. and Mrs. R. M. Booth, Miss Margaret Master George, infant, and servant, Mr. J. D. F. Eggers, Miss L. Finch, Mr. and Mrs. W. Finkernagel and child, Mr. and Mrs. Edward Turner and Miss Turner, Mrs. and Miss Rohde, Mr. and Mrs. Littlejohn and Miss Littlejohn, Mrs. Whitelaw, Miss Simpson, Miss M. II. Priestley, Miss Tellbuscher, Miss Walker, Mrs. Parkinson, Mr. and Mrs. A. Rivett, Mrs. and Miss Gordon M 'Arthur, Rev. John H. Craig and Mrs. Craig and child, Mr. and Mrs. Hotop, Miss Hotop, Mr. Donnison, Mrs. Tindal Porter and five children and nurse, Miss Sellars, Mrs. J. J. Castell and Miss Castell, Mr. and Mrs. Anderson and Miss Anderson, Mr. and Mrs. A. Lidbury, Mr. and Mrs. Higgins, Mrs. Peacock and two children.

From Sydney: For Genoa and Naples — Mr. Robert H. Barber, Mr. C. Trabattoni, Colonel Hon. E. Brownlow: And Mrs. Brownlow, maid, and man servant, Mr. and Mrs. F. von. Hessert, infant, and nurse, Alderman P. F. Hart and Mr. P. F, Hart, jun., Mr. and Alra. Weidemann, Mr. Charles J. Ohler, Mr. C. W. Koefoed, Mr. H. Thaiter, Mr. C. Fontana, Mr. Philip Schaefer, Miss F. Rod, Mr. P. Muhlenbcin.

From Sydney for Albany — Mr. Fletcher, Miss MacMahon, Mrs. Orton and two children, Mr. and Mrs. Oscar Hess and two children, Dr. and Mrs. Zillmann, Mrs. W. Richards, child, and infant, Mrs. Charles Richards, child, and infant, Mrs. Johnson and infant.

From Sydney for Adelaide— Mr. F. Schmellitscheck, Mr. John HackshaH.

From Sydney for Melbourne — Miss Maniachi, Mr. B. Finkernagcl, Mr. and Mrs. Jas. Hartlett, Master Roy Bartlett, Mr. N. M Thomson, Mr. Tindal Porter, Mr. Chamberlin, Mr. and Mrs. W. Humme, Miss Daniels, Mrs. Slade, Mr. D. B. Paterson, Mrs. Molenaar, Miss Tilbury, Mrs. Josephine Peick. Mr. F. Hickiin, Mr. G. Tessero, Mr. W. Haden, Mrs. Clara Tamm, Mrs. Inwood, Mr. Peacock.

From Melbourne to Southampton, Antwerp, and Bremen — Mr, and Mrs T. W. Jackson, Miss Ehlers, Mr. and Mrs. P. P. Labertouche, Master Keith Travers, Mrs. E.V. Birks, Misses K. T. Birks, S. Allan, Annie Boffin, Mrs. H. E. Davison and family, Mrs. R. Travers and Miss Travers, Mrs. J. M'Gee, Mr. and Mrs. F. Allan, Mr. and Mrs, Mickle and daughter, Miss D. Travers, Miss Rowan, Mr. J. H. Flack, Mr. and Mrs. W. Harper, Miss Theda Jurgens, Miss Sewell, Mrs. Richter, Miss Campbell, Miss Levien, Mr. Steinmetz, Mr. and Mrs. H. Arundel, Mr. and Mrs. D. Mitchell, Mr. and Mrs. T. D. Edwards, and family. Rev. Mr. and Mrs. J. R. Cooper, Mr. T. Cohen, Miss B. Wallace, Mr. and Mrs. Underwood and family, Mr. and Mrs. J. F. McCrea and son, Mr. Julius Heinemann, Rev. Mr. and Mrs. G. F. M. Fielding and family, Mr. H. F. Chettle, Mr. W. Wallace. Mr. and Mrs. J. Silberfeld and family, Mr. and Mrs. W. Sewell, Mr. and Mrs. Thomas Lewis Jones.

The Barbarossa arrived at Southampton:

Brisbane Courier, **Wednesday, 19 May 1897, P 5.**

COLONIAL. INTERESTS

(By Cable Message.)
ARRIVAL OF SIR HUGH NELSON.
LONDON, May 17.

Sir H. M. Nelson, Premier of Queensland, who, with Lady Nelson and family, left Sydney by the steamship Barbarossa on the 27th March, has arrive in England to take part in the Queen's Record Reign celebrations.

Western Champion and General Advertiser, **Tuesday, 15 June 1897.**

LATEST NEWS
[SPECIAL PROVINCIAL PRESS SERVICE]
Cablegrams.
SIR HUGH AND LADY NELSON.
LONDON, Thursday.

SIR HUGH and LADY NELSON have arrived in London from the Continent, and were conveyed in a Royal carriage to the Hotel Cecil.

Still no mention of James Tyson as a passenger on the Barbarossa.

The Nelson family partook of the Jubilee celebrations and had a tour of Europe before returning to Australia, taking a ship firstly to Hobart, for a meeting, then on to Sydney, then, I guess, by train or road to Brisbane:

The Morning Post, UK, Monday, 23 August 1897, P 5.

NAVAL AND MILITARY INTELLIGENCE

Sir Hugh Nelson, the Premier of Queensland, left Plymouth by the Royal Mail Steamship Matuara on Saturday, accompanied by Lady Nelson and family.

Mercury, Hobart, 12 October 1897, P 2.

SHIPPING.
ARRIVED. -October 11.

Mataura, S.S., 6,688 tons (N.Z.S. Co.), C. Milward, from London, via Cape town and Albany. Passengers-Saloon: For Melbourne - **Sir Hugh and Lady Nelson, Sir. D. J. and Misses G., J., and M. Nelson**, Mr. H. S. Dutton. Mr. Jesse Hopson, Major Somerville; 3 steerage. Agents-Macfarlane Bros, & Co.

Mercury, Hobart, Saturday, 16 October 1897, P 2

SHIPPING
SAILED. -October 15.

Oonah, S.S., 1,757 tons, W. J. Featherstone, for Sydney. Passengers- Saloon: Mrs. C. Davis and 2 children, Mrs. Gough, Mr. and Mrs. R. Brown, Mrs. Mace; **Misses Nelson**, Barratt, Mr. and Mrs. Bellaney, Dr. and Mrs. Cruese; Messrs. H. Harrison, Poulton, John Clark, Moco, Beddome, J. H. Dhu, Jolly, J. L. Hopson, Blacklow, Thornton, H. S. Dutton, **D. J. Nelson; Sir Hugh and Lady Nelson**; 5 steerage Agents-U.S.S. Co.

Evening News, Sydney, Monday, 18 October 1897, P 4.

SHIPPING.
ARRIVALS October 17

Oonah (s), 1758 tons, Captain W. J. Featherstone, from Hobart. Passengers— Mesdames C. Davies and 2 children, Gough, R. Browne, Mace, Bellamy. Couse, Misses Barrett, Clark, **Nelson (3), Sir Hugh and Lady Nelson**, Dr. Couse, Messrs. R. Brown, R. Harrison, Poulton, Bellamy, John Clark, Mace, Beddoun, J. Delue, Jolly, J. Hopson, Blacklow, Thornton, Middleton, **Nelson**, Dutton; and 29 in the steerage. F. W. Jackson, agent.

A study of Sir Hugh's travels showed he had a valet in his party when travelling to England but there was no valet on his return. It would be normal, and normally expected, that if the family took a valet with them overseas that they would return him to Australia.

Additionally, the valet would have been with them on the European trip but this does not seem to be the case.

To satisfy my curiosity further investigation is required and I found the following:

Maitland Weekly Mercury, Saturday, 5 Jun 1897, P 11.

Among the Pastoralists and Producers [BY Harold M. Mackenzie
GLEN ALPINE

Our progenitors, no doubt, were a fine, hardy band of comfort-scorning pioneers, after the type of the well-known—**James Tyson** one who still maintains that the ground is more comfortable and better for you than a spring mattress, or that, if occasion arises, you must throw yourself over a wire fence. But we have changed all that now-a-days, and the man who still prefers a saddle and the ground, not to mention myriads of biting insects, is simply an ass. I shall be pleased to hear what **Mr. Tyson thinks of the Hotel Cecil, in London.**

Sunday Times, Sydney, Sunday, 4 July 1897, P 5.

PERSONAL

A private letter from London mentions '**Jimmy Tyson**' the squatter millionaire, who is visiting the Old Country after an absence of forty years. Our correspondent says **that Tyson is going about a good deal, and as a result of free association with his fellows, will return a changed man.** It is to be hoped he will be a more liberal one.

Western Grazier, Saturday, 10 July 1897, P 3.

BREVITIES

Millionaire **Tyson,** the cattle king of Australia, was mentioned as a prospective record reign knight, but he is not listed. By the way, **"Jimmy"** is now in London, no end of a swell, and is "going it" regardless of cost. He will return a wiser, but improbably a poorer man.

Well! Did James arrive in England? I tend to think he did. The reports from apparent different sources suggests he was there, but no reports from the city newspapers verify it.

But how did he get there and return?

The Nelson party included a valet when boarding the Barbarossa but there was no valet when they returned to Australia. That should never happen. If a person is employed as a valet and is taken overseas that employer should return with the valet. I believe the valet was James. I also believe the city reporters knew James was in England, possibly through their British contacts, but I do not know if they suspect the valet was James. In the passenger listing I notice the valet appears to be with the Premier's son rather than the Premier on the passenger list, possibly to shield James from the reporters. It could be that James' presence in England was first announced to the Australian press by the British press. I do not understand why the provincial press made reference to it but the city press did not. I found no evidence that James' family knew he went to England.

I don't know how James returned to Australia but for James, who said he travelled second class on Australian trains "because there is no third class," would have no qualms in travelling in steerage. And, of course, steerage passengers are not named so James could have returned any time anonymously

Brisbane Courier, Tuesday, 11 August 1891, P 2.

A WELL-KNOWN QUEENSLANDER

Meeting a friend on one occasion on the platform at the Orange railway station, the friend expressed surprise at seeing Mr. Tyson riding in a second-class carriage.

"Do you know why I do ride in a second-class compartment?" said Mr. Tyson.

"No, I do not know why," said the acquaintance.

"Well," said "Jimmy," "it is because there is no third-class," and with a broad smile he resumed his seat, and the friend looked crestfallen, and drowned his contempt for the " old fellow" in a bottle of Bass ale.

The statement, "after an absence of forty years" had me scurrying back to NLA Trove and this is what I found:

Empire, Saturday, 14 May 1864, P 5.

DEPARTURE OF AN OLD COLONIST –

We hear that Mr. James Tyson, of the Lachlan, one of the most extensive squatters in the colony of New South Wales, has just left us on a three years' tour - first to Van Dieman's Land, and then to England. Mr. Tyson was a bona fide squatter, who has worked his way up from a very humble position until he has become one of the richest men in these colonies.

He was a thorough practical man, and during his career he went through an amount of hardship and toil in the interior seldom gone through by many of his class. Although bearing the reputation of being an uncommonly hard dealer, he was conscientiously scrupulous in the discharge of his just obligations in this sense he was extremely correct.

He was a liberal supporter of all local charities. We wish him a bon voyage, and trust that be may return to the land of his birth in good health, and satisfied with his tour through Europe. Mr. Tyson is unmarried, and has left behind his brother, Mr. P. Tyson, to manage his vast squatting properly held by the former in New South Wales

.

We may never know, for sure, if James did go to England in 1897 or 1864; how did he get there and back, who did he see and how long was the visit. New information may come forward one day. Meanwhile, James died taking the answers with him leaving the accepted view by most people that he never went overseas.

Similarly, James went to New Zealand but apparently none of his family knew and, again, it was reported in the provincial papers but not in the city papers.

Albury Banner, Friday, 18 Apr 1884, P 15.

A RIVERINA MILLIONAIRE.

Mr. James Tyson, the Australian millionaire, is (says a Wellington paper) paying a visit to New Zealand. Sir William Clarke, Sir James Elder, Sir Samuel Wilson, are all Australians noted for their wealth, and justly so, for they are wool kings indeed. But all these and other landowners must bow their heads to Mr. Tyson. Three years ago, the London World ranked him among the very wealthiest men in the world. He is more solidly wealthy than several of the so-called wealthiest men, for he has no liabilities He is the greatest sheep owner in the world, and his many herds of cattle have a great reputation. It is said he owns 1,000,000 cattle. In addition, he lends more money on mortgage than any man in this hemisphere.

It is understood that Mr. Tyson is looking out for some properties worthy his attention in New Zealand, and is now in the Wairarapa district.

Maryborough Chronicle, Wednesday, 9 April 1924, P 12.

PIONEERS OF AUSTRALIA.
INTERESTING REMINISCENCES. THE LATE MR. JAMES TYSON.
By Joseph Stirling

Among Australian pioneers there was no one so thoroughly misunderstood as James Tyson. He had the reputation of being a multimillionaire. When I first saw him, he had previously bought Felton station on the Darling Downs and some Victorian acquaintance of his gave him the name of "Hungry Tyson."

I had been in Toowoomba some years when a man of rather forbidding appearance came into my blacksmith's shop and asked the price of a newly finished dray. When I told the price was £12/10/- he said he could get a dray like it in Melbourne for £10. Eventually he offered £11, but there was no reduction in price, so he left in dudgeon. When he was gone a bushman said, "That is 'Hungry Tyson' and there was no likelihood of doing business with him.

I thought I knew him when he called a few days after and asked the price of a pair of heavy wheels. Not leaving the anvil, I said, "It all depends on size and weight of tyres." Having explained, I told him the wheels he wanted would cost £10 per pair. He said he would give that price and asked who would be the maker. There were several wheelers there at work and I pointed out Pat Dunne. Tyson said to Pat, "You are going to make a pair of wheels for me and I expect you to make them well, as I am giving you ten shillings extra for yourself. I must see them after they are tyred and before they are painted. "Of course, Pat had licked the blarney stone and gave a sample. There were six or seven pairs of wheels to same pattern with some payment to Pat. We got acquainted. Mr Tyson was proud of his own ability in the smithing line and told of his early training in a smith's shop where he got a weekly wage of 15s. His ability in that line enabled him to cut a set of tyres several times in the back country. **One story led to others. He told us he had been travelling in New Zealand, where the grandest timber and best timber machinery in the world was. Logs of great size were hauled like whip handles. He was travelling as John Smith., but somehow it got out that his name was Tyson and, owing to that he could buy nothing in New Zealand. He visited a native residence where there were many families with 25 children. There he saw a fine buxom young Maori, a chief's daughter, who had £13,000 a year in her own right. New Zealanders have not been like Australians who allowed their claims on land to lapse.**

The New Zealanders are increasing and look after their tribal rights.

There were many among the new Zealanders who looked very big men, but Tyson was a big man too, measuring six feet two- or three-quarter inches in height and forty-eight inches around the chest. At the encampment many of the Maoris thought they were bigger than Tyson but the tape is a strong argument. Looking to buy land in New Zealand, Mr Tyson found the best farms were all owned by Scots or their descendants. Eventually, the knowledge that his name was Tyson hunted him as a buyer from New Zealand (sic).

At his invitation. I often visited Felton, Tyson's home and he took a delight in pointing out his own improvements. The fences were much stronger than those of the usual stations. Some of the great favorited grasses were peculiar to Felton. Several graziers in the locality had collections of the best grasses and seeds had for some years, and at proper seasons, been sewn in Felton paddocks. In regard to stockyards, I remarked the gate hinges were surely too heavily made of 2½ inch iron. "You see," said Tyson. "When a bullock charges that gate the bullock is dead." that seemed to him a complete answer. At one time I went on to the head station and found only the cook, a young woman, an antiquated old bushman of 75 and one cat. Telling Mr Tyson, the circumstances and the date, he seemed much pleased. On that date there were eight men and four dogs at work on the station. Every man and every dog had a day's work on Felton. There was only one woman employed — the cook-- on the station. Neither superintendent nor any other married man had wife on the station. Shortly after, I was speaking to Archie Meston as to peculiarities of squatters, and he had quite a string of these peculiarities to recount.

CHAPTER 8

RUMOUR: James knew his mother was a convict

It is now well known that James' mother, Isabella, was a convict who was banished to Australia for seven years after being found guilty of minor theft. This is not unusual as England wanted to get rid of its citizens found guilty of a misdemeanour by transportation to land far away or through the gallows.

Isabella was fortunate because her husband, William, accompanied her. When they arrived in Australia, Isabella was assigned to William so they were able to establish themselves as a family. Their first-born child, Margaret, only three years old, was left in England with grandparents and never heard of by this family again. Their second child, William II, a baby of one born in Newgate jail, would have had no memory of Isabella's traumatic circumstances before arriving in Australia.

As William accepted responsibility for Isabella there were no obvious restraints on her. The seas around Australia and huge distance from England was enough satisfaction by the British that they were rid of their miscreants. The small Tyson family, to become very large over the years, could live like free settlers and this is apparently, what William and Isabella told their children.

It appears only officials who worked with convict records would have known Isabella was a convict; and, then, it would have been an entry in a very large book. So, even fewer officials would have been able to put a face to the entry for Isabella Tyson. But the British authorities wanted to keep a check on the people in the penal colony of Australia, consequently, regular checks on the population were conducted.

About every four years there was a muster (or census) of all residents not incarcerated, a chore usually performed by the head of the household at a nominated place. While he was alive William would have performed that duty and after he died in 1827, Isabella (as she got her Ticket of Leave in 1825) would have been the household head. That entry would have nominated Isabella convict or ex-convict; an item not seen or discussed with James or his siblings.

It is hard to imagine Isabella was transported to Australia in 1809 and it was not known by her children and their progenies for at least 89 years. It is assumed the truth came to light when James died.

Consider the following letters, an entry in the authoritative *Australian Dictionary of Dates and Men of the Time* by J. Henniker Heaton and a report by a sister.

Letter to Geo Tyson 18860703 Transcription.

> Felton Station, July 3, 1886
> Via Cambooya
> Darling Downs
> Queensland
> Australia
>
> Geo Tyson Esq
>   112 Olinda Road
>   London N.
>
>   Dear Sir,
>   I am receipt of yours dated 26 March last (1886) and making enquiries as to the probability of you being related to me. In reply, I cannot see how such can be the case when knowing the line of descent as given by you. Your relatives appear to have sprung from

the neighbourhood of Lancashire. I notice that you say about an uncle of your father named William Tyson having migrated from Lancashire on or about the year 1818 or 1819 with 2 sons and their wives & you think they were bound for America. Strange to say that a person who had lately come from America on business, but have forgotten his name, but who is now living in Sydney who married a Miss Tyson one of a family of the name in America who I was invited to see, but did not do so. She was descended from a William Tyson and no doubt is the direct issue of one of the Tysons you speak of as having emigrated to America long ago. There is also a family of the name in I think the River Hastings in New South Wales, who claimed relationship to me some time ago but they were wrong. There is another family of the name at or near Taradale in Victoria. There is also a Mr Thomas Tyson Merchant of Elizabeth St., Melbourne who is agent for the Carlisle Ale and there are others very thinly sprinkled through Australia, none of whom have I ever seen, but I understand they are mostly from or descendants of Cumberland. This is evidently the headquarters of the Tysons. This I have learned from several parties who have been there. I also had a small paragraph sent to me by a friend who has been in Cumberland in which it was said that if Her Majesty required strong able men to defend her, she could find such amongst the Tysons of Cumberland. Thus, it would appear they are very numerous at what is supposed to be the headquarters of the families of that name. Whether all those who bear the name are descended from the Gilbert Tyson you speak of as one of the followers of William the conqueror or not, it is hard to say, but it is strange that all of the name or the offspring of the name appear to have originally come from that or near the place named.

As far as I am personally concerned, my father came from Cumberland & through some family short comings enlisted & went into the army. My mother was from Whitehaven or Newcastle. My father bought his discharge & came out to Australia I think in 1808. He accompanied a gentleman who came out to inquire into charges brought against the Governor Blythe in New South Wales. He then went farming and died when I was only 6 years old. Thus, I know but little of my ancestry beyond what I have here stated.

In conclusion, the delay in replying to your letter now under notice was owing to its having been by some means having been mislaid for which I have to apologise and in further concluding I think it is likely I may visit the old country soon, and if I do, I will not fail to call on you, if for nothing else it will perhaps afford each of us pleasure to shake hands with namesakes, who are at any rate very scarce in Australia.

I am Dear Sir
Yours very
Faithfully,
James Tyson.

The other letter was received by a claimed relative in England reported in the press after James' death but received 10 years prior: -

Bristol Mercury and Daily Post, Wednesday, 28 December 1898, Issue 15797.

THE TYSON MILLIONS
Interesting Letter

It has transpired that there is a resident in Ramsden Street, Huddersfield, a relative of the late Mr. James Tyson, the Australian millionaire, the destiny of whose riches has lately been the subject of much interesting speculation. The relative mentioned possesses a letter which he received from his kinsman ten years ago, and this sheds some light on the identity of the wealthy squatter. In the letter Mr Tyson said:-

As to myself personally I dare say you have heard a great deal, a little of what may be true, but in speaking of myself it should be admitted that I know something, and in saying this I really admit there is not much in me. I am a simple, unassuming bushman, born and bred without much education. The only thing I feel I have a right to be proud of is my simplicity, sobriety, strength, health and steadiness in pursuit, and unwearied application.

As to my parentage I know little. My father came from Cumberland or Northumberland, and enlisted in the army, and was put in charge of a beacon with a staff

of men to make an alarm of fire on a hill in case an alarm was given of the approach of the French fleet under the late Napoleon Bonaparte. After this my father bought his discharge, I am told, and came out to Australia with a government officer, who came to make enquiries into charges against the Governor of New South Wales in or about the year 1808. My father died when I was about four or five years old, and thus I know little, only that my mother said that he came of a very good family. My late brother, William, was born in Yorkshire[16], and an elder sister, Margaret, was left with his (her) grandmother. All my other brothers and sisters were born in New South Wales. They are now nearly all dead, but have left numerous offspring both on the male and female side.

In conclusion I am still thinking of visiting the land of my fathers, if only for a day or two, and, if I do, it is likely I may call on you to fraternise and shake hands. I get a good many letters from people of my name in Darwen who claim relationship, which they seem to be found on the emigration of a Mr. John Tyson only 30 to 40 years ago. There are several Tysons in Australia, although I have not seen them. Some of them are located on the Nambucca River in New South Wales.

For your information as to myself I send and article taken from the "Town and Country Journal" of New South Wales, extracted from the Australian history of "Men of the Times," which is fairly true, but the year 1818 is used instead of 1808, and in the matter of cooking the duck, while they are a favourite sport and welcome grill, I never use anything other than salt.

The visit to England hinted at was never made.

---

[16] It is now accepted William II was born in Newgate Prison

I found the entry in "Town and Country Journal" mentioned in the previous letter and I reproduce it in full as James said it is "fairly true." This is important in this exercise as it "fairly" validates what is written as a summary of James' life.

The article does not acknowledge the source but I suggest James contacted the newspaper for that. He was told or remembered was told The Australian History of Men of The Times which could only be Australian Dictionary of Dates and Men of the Time by J. Henniker Heaton as it has the exact wording.

Australian Town and Country Journal, Saturday, 8 Mar 1879, P 17.

Mr. James Tyson-The Australian Millionaire.

James Tyson, the well-known Australian millionaire, was born at Cowpasture, near Sydney, on April 11, 1823. His father, William Tyson, was the scion of a good old Cumberland family, but, having offended his parents by marriage against their wishes, he found things so unpleasant at home that he enlisted in the army. His discharge was purchased about 1818, when he emigrated from England in the service of Mr. Commissioner Bigge, who was sent out to investigate the charges that had been made against Governor Macquarie. Mr. William Tyson was kept by Mr. Bigge some time in his service at Government House, Sydney, and was asked by Mr. Bigge to accompany him to India, but, having a son (the late William Tyson, of Geramy), Mrs. Tyson objected to go to India, thinking the climate would be prejudicial to the child. Mr. William Tyson then commenced farming near Baulkham Hills, and afterwards received a grant of a farm near the Cowpastures, and where his son James was born. Mr. William Tyson did not succeed very well with his farm, and he received the grant of another at East Bargo, where he died. After assisting his mother some time on the farm at East Bargo, James Tyson entered the service of

Messrs. Vine, at Brook's Point, near D'Arrietta's Farm (near Douglas Park) as working overseer, at a salary of £30 per annum. He afterwards transferred his services, in the same capacity and at the same salary, to the late John Buckland, Esq., of the Oven River. His next step was to a similar situation, with a rise to £30 per annum, at Jugiong, with the late Henry O'Brien, Esq., of Douro, near Yass, from here he went to the same gentleman's stations at Groongal, on the Lower Murrumbidgee, and remained there till he joined his brother William in the formation of a station called Gunambil, on the Billabong. After putting up a hut, yard, and paddock, the task devolved upon James of going to Burragorang for a draught of cattle, which Mr. Graham of Campbelltown had agreed to place in the hands of the brothers Tyson. James Tyson, to prepare for the journey, cooked as much rations as he could carry on his horse, and of money he had just one shilling, which when he reached Gundagai was demanded of him by the punt man for ferrying him and his horse over the Murrumbidgee. Thinking he might want the shilling for a still greater need, Mr. Tyson determined to save it, and, declining to use the ferry, swam over the river, if not at the risk of his life, at any rate greatly to the detriment of his rations. After numerous shifts and difficulties, Mr. Tyson got the cattle together, and drove them as far as the Murrumbidgee, where he met his brother, who had been compelled to abandon the newly formed Gunambil Station on account of the water having utterly failed and who had sold the run and improvements for £12, but—did not get the money! They then went to the stations now held by James, near the junction of the Lachlan and the Murrumbidgee Rivers. Whilst his brother William carried on a dairy, James went jobbing and cattle droving, until a few of his stock were fat and fit for market. He then joined with the neighbouring stockowners, and made up a mob for Sydney, selling his first lot to Mr, Thomas Sullivan (now of Sullivan and Simpson) at £3 a head for the pick and £2 for the remainder, whilst the same buyer purchased a lot from the Murrumbidgee at eight shillings a head, which were afterwards sold at 6d. profit to a Mr. Inches for boiling down purposes. The run near the junction of the Lachlan and the Murrumbidgee was taken up by the Tyson brothers, July 8, 1846, and was held by them for about four years without a license, the Government having refused to grant licenses for the runs on the north side of the Lachlan, as no Commissioner of Crown Lands had been appointed for that district. The runs were afterwards thrown open for tender, and the Tysons sent in one which was not accepted. The Tysons, however, purchased the right of lease from Mr. Flood, who was the successful tenderer, and so remained in undisturbed possession of the Towong or 'Tysons' Run.' They also held a licensed run on the south side of the river opposite Towong (Toorong); and when the brothers dissolved partnership, Mr. William Tyson took the run on the south side, and Mr. James Tyson that on the north, and it has ever since remained in his possession. In 1851, when the gold discoveries were made, James Tyson commenced cattle-droving to Sandhurst, where he opened a wholesale and retail butchering business, and where he made large sums of money. After carrying on business successfully at Sandhurst until 1855, Mr. Tyson purchased the Royal Bank Station near Deniliquin ; he afterwards purchased the Juanbong (sic) (Juanbung) and other stations on the Murrumbidgee, then the famous Heyfield Station in Gippsland; he next extended his operations to Queensland, where he purchased the Felton Station on the Darling Downs; he afterwards acquired several immense stations on the Warrego, where, as in Victoria and this colony, he now holds large areas of freehold lands. Mr. Tyson was a broad-shouldered, robust man, standing 6 feet 3½ inches. He has never had a day's illness in his life; has lived much in the open air, and prefers it; was a keen sportsman and a good shot. He was a true friend and staunch protector of the Aboriginals on his various stations, who were all very much attached to him, and rendered willing service. He was of a very retiring disposition, and always refused to allow parliamentary or other public honours to be thrust upon him. He was a bachelor, and mingled but little in society; was, however, very fond of children, and had always been a liberal supporter of all local schools, and also a liberal subscriber to all local hospitals and other popular institutions, although generally desirous to avoid having his name paraded before the public. The amount of Mr. Tyson's wealth cannot be easily estimated, but it may be mentioned that at one time he was able to offer the Government of Queensland a loan of half a million of money towards the construction of a proposed transcontinental railway. Mr. Tyson owes his good fortune mainly to his energy, his untiring industry, and his great self-denial. He never indulged in a glass of wine or spirits or in tobacco in his life, and those who know him best say, as Disraeli said of Gladstone, that he had not "one redeeming." his temper was so even that under the most trying circumstances no profane word had been heard to escape from his lips; and the frugality and simplicity of his habits should disarm the envy of those who might be disposed to covet his great riches.

It is interesting that James corrected the date his father left his military service but not the date of his birth. He was not born on April 11, 1823 but April 8, 1819 as shown on his baptismal certificate (T.Y.S.O.N. by Zita Denholm Page 21.)

A Briton wrote to the executors of the estate claiming relationship, hoping to get a share. Their argument was so convincing a hearing was held in March 1903. Details of their family were irrelevant because they were not successful so I have not reproduced them. However, the comments by the "master" in the decision suggest at that time it was still not generally known James' mother was a convict.

The Australian Star, Sydney, 5 March 1903, P 6.

TYSON MILLIONS
An English Claimant
JUDGMENT IN EQUITY

...

The Master quoted an article from the "Northern Daily Mail" of August 28, 1889, headed "James Tyson, the Millionaire; his history related by himself." In this there appeared the following extract from "Men of the Time" respecting the father of the Intestate. "His discharge (from the Cumberland Militia) was purchased about 1808, when he emigrated from England in the service of Mr. Commissioner Bigge, who was sent out to investigate charges that had been made against Governor Macquarie." On this the Master commented that the evidence in this case had proved most conclusively that William Tyson, Isabella his wife, arid their son William, came to Australia in the ship Indispensable (sic), which arrived in Sydney on August 19, 1809, amongst the passengers being one Mr. Hartley, R.N. There was also evidence that Mr. J. T. Bigge arrived in Sydney in the ship John Barry on September 25, 1819, and therefore, as the intestate pointed out in the letter to W. H. Tyson, the article in "Men of the Time" was erroneous, insofar as it stated that his father, William Tyson,' came to the colony with Commissioner Bigge. That error was evidently unknown to the witness, who had adhered too closely to the published history of the intestate, and had thus put into the mouth of Betty Coward a declaration as to the fact that they now knew had never been, except as the creation of the edition of "Men of the Time." After, that expose, the Master said, he need hardly say he attached no weight to the evidence of the witness.

On the evidence of Mrs. Charlotte Heap, the Master cited the evidence she gave in cross-examination as to letters she had written to Mrs. Clarke, in which she said she thought William Tyson had been bought out. The records of the War Office showed the court that he was discharged from the militia on July 19, 1809, his term having expired. That afforded strong proof, said the Master, that the witness was biased (sic) by the newspaper article. The plaintiffs in Tyson v. Doneley had gleaned facts from intestate's private papers which had never become known to the witnesses in England.

An extract, from an autobiographic entry in testator's pocket diary, is given as follows, under date December 28, 1867: —

"Father's mother's name was Margaret."

Now, if that were correct the intestate's father could not have been a son of Nicholas and Esther Tyson. In a letter to George Steel the Intestate wrote: — "My sister Margaret, who was left with her grandmother, I am told, went to Scotland." If that were correct It was inconsistent with claimant's theory that testator's " sister went to Liverpool and married and died there. In 1848 an application was moved by the intestate's eldest brother as heir-at-law of their father to the Court of Claims in this State and from the evidence of the mother of intestate in support of that application it would appear she believed her eldest daughter Margaret was then alive. This would discount the assumption that her eldest daughter Margaret was identical with Margaret Fish, whose interment on October 14, 1834, was registered. For those reasons, the Master said he disallowed the claim. It, therefore, became unnecessary to consider whether claimant was the daughter of William and Margaret Fish.

His Honour, after hearing argument, dismissed the application with costs.

The proven history of James Tyson's family in Australia as detailed by his biographer is as follows:

> **T.Y.S.O.N., by Zita Denholm, P 23.**
>
> The migration of the Tysons to Australia was not voluntary. Isabella and William Tyson and the then baby William (it is easiest to call him William II) arrived in Sydney on 19 August 1809 in the *Indispensible*. Isabella was one of sixty-one female prisoners on board. William had obtained a berth in the service of Mr. Heartley RN, one of the passengers.
>
> When they disembarked Isabella was assigned as the housekeeper to her husband. She had already served sixteen months of her sentence of seven years' transportation for theft before she set foot in Australia. William Tyson, in spite of four years' service in the Royal Cumberland Militia Grenadiers in the late years of the wars with Napoleon, was no townsman. All the family tradition points to his having been raised on a farm, and shortly after his arrival he commenced farming at Baulkham Hills.

There was a muster of the population every few years, which are completed by the head of the household who was to nominate the people living in the household. The status of each person is listed and, if they came from overseas, the name of the ship is to be noted. So, James' father would have completed the census until he died in 1827, when his mother would have taken over that duty detailing that she was a convict without telling her children.

The colonial authorities are quite happy for a man to accept responsibility for a convict, who effectively becomes whatever the "master" wanted from a slave to a soul mate. Obviously, James' father chose the latter as he left everything in England to accompany her to Australia.

It seems that James and, maybe, all his siblings before him died without knowing that their mother was transported to Australia as a convict.

CHAPTER 9

RUMOUR: Tinnenburra woolshed was the largest in Australia

Many stories circulated around Australia since the Tinnenburra Woolshed was built in 1893. Variously it was called: largest in Australia, largest in the world, over 100 stands ("stand" is an area for a shearer) to name a few. To put the record straight it had 60 stands. I attach the description of the construction which goes with the construction plans in Appendix 1.

Sydney Mail, Saturday, 30 December 1893, P 1369.
A Model Woolshed.
By Wynne Gray

The Hon. James Tyson's new woolshed erected at his Tinnenburra Station, South-western Queensland, is built on the site of one of the artesian bores situated between the Tuen Creek and the Warrego River, and lies about 18 miles north of the New South Wales border township of Barringun. That this shed is one of the most perfect in the Australasian colonies, a careful perusal of its description will prove, and a.

personal inspection of the building itself will demonstrate to the most carping critic the excellent care that has been exercised in its construction by the architect and builder, Mr. Timothy Hannay, who has had entire charge from start to finish.

The shed stands upon 1410 blocks, which are 7ft. above ground, and of an average depth of 2ft. 6in. in the ground. No block is of less diameter than 10in. at the small end. These were first erected perfectly straight in lines, rammed firmly, and afterwards sawn-off level. They are all upon solid bottom and require no packing, which is very essential in making first-class work. As the sleepers used throughout are 8in. by 6in., and are well spiked to the blocks with 3/8in. iron spikes, there is strength enough upon this foundation to bear any weight.

The interior of the shed is constructed upon what is known as the "Centre Floor System." This is a most convenient form, as the overseer of the "board" has all the men in view at one and the same time. It is built on the T pattern, the front, being the wool press and classing-room covering 10,000 square feet of space, is capable of storing 3000 bales of wool. Two presses work, one on each side of the main entrance, and opposite to each press is a platform built to the floor level at which four wagons can load at once. On the platform at the main entrance is an office where all the clerical work in connection with the shed is transacted. The distance between the joists in this part of the shed is only 10in. The flooring is shot, and the boards measure 4in. by 1in. This makes a very strong job, most necessary where it is required to store heavy weights. This building is well lighted by "attics" or "dormers" in the roof, besides "lights" between the skillion and main roof, and is intersected by the main building in the usual manner. The main building covers 21,000ft. of ground space, of which the shearing floor is part. The shearing floor measures 150ft. in length and 26ft. 4in. in width. There are 30 stands on each side of the floor, and a tram line in the centre to convey the wool, to the tables. The catching pens measure 11ft, by 10ft., and the partitions are constructed of bar iron with heads and nuts. Each shearer has his trap and shoot to carry his shorn sheep to his yard outside. Also, each man has his own door leading into the catching pen, which door shuts in and out by means of springs and closes automatically. A 6ft. 6in. wall shuts off the shearers from the sheep so that in penning-up the sheep cannot see anything outside the catching pens. By this means the sheep pen-up with much less trouble than when they can see beyond their pens. This wall is constructed half its height with corrugated iron, and the remainder with louvres. Behind the pens is the feed race, and the catching-pen gates open to this race fence and also close the pen. Outside this race is a main race in which the penners-up work. The fence dividing these races is of wire, treble plaited, and strained with eyebolts. The sheep-storage room is subdivided into races all running parallel to one another. The entrance at the yard end of the shed, has a double race on an inclined plane. All fences and divisions are of treble-plaited wire with a. 4in. by 3in. cap, and posts 4in. by 4in., 2ft. 6in. apart. This building has an opening covered by an awning 6ft. wide which gives ample light the full length and both sides of the building. Through the entire structure there is also, a line of glass sash lights, 2ft. wide, running the whole length on both sides between the skillions and main roof.

The main walls are 15ft. high from the floor, and the mam building posts are 8in. by 8in. The walls are braced longitudinally with 6in. by 2in. braces tenoned, and draw-pinned with ¾in. iron pins. The tie beams, every 10ft, are angle-braced to the main posts with 6in. by 3in. braces.

The roof is constructed with principals, one every 10ft., and is made as follows — Sole plates, 6in. by 4in; rafters, 6in. by 4in. collar tie, 6in. by 4in.; king post, 6in. by 4in., diminished to 4in. by 4in.; queen post, same as king post struts, 4in. by 4in., straining struts, 4in. by 4in.; the whole is tenoned and draw-pinned, and at every joint are ¼in. iron plates set on top of the tie beams and bolted to them.

The ends of the sole plates project 2ft, beyond each side of the building, and strong anchor bolts are fixed to the building posts, and bolted through the sole plates. There are no nails used in the frame work of the roof, every joint being bolted. The walls throughout are of iron nailed on; the roof iron is screwed on. It has taken 27cwt. of lead-headed nails to affix the iron, and 42 tons of iron to cover and complete the shed. Two tons of sheet lead have been used for valleys, &c. Besides these, 2 tons of nails have been used, and the same weight in bolts of various kinds. The plates and straps alone used for fitting the principals totalled in weight 2¼ tons. Two hundred pairs of gate hinges and hooks were manufactured for the shed at Sydney. All entrance doors are framed; some run on wheels, and others are raised with pulley weights.

The sheep-storage capacity on the floor is 8000, and as many more can be stored beneath if required. This shed is built for machines, and, should they be erected, it is the intention of the manager to utilise the flow of water from the bore as a motive power, the bore being in the shed yard and yielding a large supply.

The total quantity of timber used in the erection of the shed was about 300,000ft., the whole of which was cut on the ground by the station plant. The time occupied in completing the building was seven months. It was built on the piece-work system throughout, an average of 10 carpenters being employed.

The huts for shearers and rouseabouts are a new departure; the dining rooms on the ground floor, and the sleeping rooms above. The boards of the sleeping-room floors are matched in order to prevent any dirt from falling on the tables below. These huts are economical as well as clean and comfortable, as both rooms are under one roof, and, the sleeping and eating rooms being separate, they form a very comfortable contrast to the old style of building.

The builder of this shed, Mr. Timothy Hannay estimates the cost at about £10,000. I have, however, just seen Mr. James Tyson, who informs me that the cost has not exceeded £7000. The builder is his own architect, and has worked out and furnished his own plans. He was entrusted with charge of laying out the work by the proprietor, Hon. James Tyson, and has given the greatest satisfaction.

The accompanying drawings will explain very clearly the general construction of the shed and yards, but I should advise all those who may happen to be in the vicinity, of Tinnenburra Station to call and view for themselves one of the finest woolsheds in the Southern Hemisphere.

CHAPTER 10

RUMOUR: James Tyson was a poet

Due to items of poetry found in James Tyson's diaries some people believed James to be a poet. Unfortunately, I am dispelling that rumour.

Sydney Stock and Station Journal, **Friday, 8 December 1899, P 10.**

Tyson as a Poet.

Many hard things have been said about the late James Tyson, who was probably as good as his traducers, if not better! He had his faults, no doubt, but if only the faultless ones said hard things about him there would be little evil spoken about James. It seems a pity to give currency to a new charge against a dead man, but this is such a novel one that it is interesting. He is accused of having been a poet! It does not bear the mark of credibility on its face, yet you never can tell! Even the strongest men have their moments of weakness, and poor James may have yielded in a fit of temporary aberration. Still, there have been poets who were good men, and it is not fair to say that all poets are alike.

Here is a memorandum that reached our office the other day,

through a squatter who has a soul to respond to such things: —

Copied from James Tyson's pocket book, found at Heyfield, October 2. 1899, supposed to be his own composition.

Ingremba, 16 Nov., '63

Let the wealthy have their customs,

And keep the wealth they own;

But, while claiming all their pleasures,

Let other men's alone.

Let them boast their halls and mansions.

And live as best they can,

For we have learnt 'tis love and home

That makes the richest man.

7 June, 1864 —

Made a wager with Mr. Rutherford of the best new hat that James Tyson is not married within 2 years.

John Adams, Witness.

Perhaps poor James yearned for love and home, and looked with envious eyes at the little huts of his own boundary riders, where the happy laughter of childhood was ringing. You never can tell. He didn't get married, poor chap! I wonder if Mr. Rutherford got the hat?

The poem listed above came from a poem titled, "Let the Monarch Boast his Title" by Robert Rutherford published in a 1857 in a book Titled, "The Farmer's Boy."

CHAPTER 11

RUMOUR: James Tyson never had any formal schooling

Family rumour was that the only schooling James and his siblings received was from Thomas Clements.

Thomas Clements was a convict transported to Australia for fourteen years for forgery. He arrived on 28 July 1814 and was subsequently assigned as servant to James' father, William. Whilst James' parents were illiterate (at least when they came to Australia), Thomas was not. When James was seven his father died and Thomas married James' mother.

The following two articles suggest that James had at least two years schooling in a formal class-room.

The Sydney Morning Herald, Saturday, 12 November 1938, P 21.
Appin Village
Where History Was Made.
By F. Walker, F.R.A.H.S.

...

Two notable men who lived in the Appin district in the early days were James Tyson, who died in Queensland in 1898, and Hamilton Hume the noted explorer. The Tyson family lived at Wilton, a few miles from

Appin. James Tyson went to the old school at Appin and was educated by Mr. Joll. "No one thought," said an old resident, "that he would ever become the great and rich man he did, for he used to run about this district bare-footed, just a careless thoughtless kind of boy." Hamilton Hume lived at Rockwood, which is about midway between Appin and Campbelltown. A memorial tablet near the site of his old residence was elected by the Historical Society. in 1924 the material being largely composed of the stones from his old home nearby.

Queensland Times, **Tuesday, 6 December 1898, P 2.**

Our Brisbane Letter.

[From Our Special Correspondent.]

Monday, December 5.

THE death of the Queensland millionaire is filling the papers with Tyson stories and Tyson sayings. It is expected to fill a corner of the Treasury chest, too. No one knows—not even himself knew, as I will show presently—what his estate is worth, but it will be a round sum; and, as the succession duties increase with the distance of relationship between the demise and demises, they will operate to some tune in this instance, seeing that nephews and nieces are the nearest relatives of the deceased millionaire that remain. Mr. Tyson Doneley, son of a sister (Margaret, I think), of Brookstead, Darling Downs, is likely to drop in for a "pot." About 18 months ago, as I think, I sought the Hon. James Tyson with a proposal that I should write his life—he to tell it to me as occasion served. But the old gentleman was "not on," and, knowing the futility of attempting to change a resolve of his, I did not attempt it. The reason alleged for his non-compliance was want of time. He would never be able to find the time necessary for the interviews. But before that interview—which lasted quite an hour—was over, I got my own notion of what was the matter, "I often thought," said Mr. Tyson, "that if I broke my leg-met with an accident that laid me up for a good while—I would write my life myself. I have it all here," he said, producing a bulky notebook from his breast pocket. I wondered to myself whether the old fellow had come to the conclusion that there was money in it, and therefore that it was too good a thing to give away to anyone. The Hon. gentleman, having first "shouted" -aye, "shouted"—he took lemonade, or some such innocent beverage, himself—proceeded to talk with the most charming frankness about anything I led up to. You may be sure it was about himself. **I see a statement in a paper today that " the young Tysons were well educated." "Well educated" is a comparative term. " I went for two years to a school about five miles from my father's place," said the late millionaire. " Was that all the education you got, Mr. Tyson?" "There was an old soldier living with us who taught us before that," was the quiet answer.** The papers are already quoting favourite expressions of his. Here is a very favourite one: "My methods are very simple, and I am a very simple fellow myself--I've always said so." And he told me how he worked his banking and station accounts to support each other " feed" each other, as he put it. He did not know exactly what his balance in the different banks came to, but it was " some hundreds of thousands of pounds." " But did you never have the curiosity, Mr. Tyson, to make out an estimate of your different properties, to see what you were actually worth?" I asked, naturally curious on that question. "No," he said, laughingly, " I have never done that." And so, we have had to wait for death to solve the enigma. The whole colony is pecuniarily interested in it now.

The Flora Shaw's comments show how well James had apparently taught himself on subjects most people cannot even comprehend, then hold discussions on the subjects is amazing. See her comments below.

Clarence and Richmond Examiner, Saturday, 10 December 1898, P 4.

JAMES TYSON.

Several months ago, one of the late Mr. Tyson's bankers told the writer that the great pastoralists estate was worth at least £6,000,000, and that he had not made a will. His opinion was that he would not make a will. This opinion seems to be confirmed by news received from Queensland to-day, which is to the effect that no will has yet been produced. Tyson appears to have had as great a prejudice against making a will as he had against calling in a doctor when he was out of health. A strong-willed man, he thought he could overcome the disease that killed him without the assistance of a medical man; but he went down in the fight. And yet he was supposed to possess more than an ordinary share of common sense. Few men, however, have sufficient common sense to apply to all the affairs of life, and so the majority are often be fooled in one direction or the other. If it should be proved that Mr. Tyson left no will, his relations will doubtless be glad, because the probability is they will get more out of the deceased man's estate than they would have done had he disposed of his property in the usual manner. **It will be remembered by some of your readers that a few years ago Miss Flora Shaw visited this colony as the representative of the London Times. She had heard a good deal about Mr. Tyson, and formed the opinion that he was rough, uncultured and uncouth. As it happened, however, she spent several days in his company in the Hay** district, when she discovered to her astonishment, that his reading had been nearly as wide as her own, and that he knew a great deal more about the works of the modern philosophers and scientists, such as Herbert Spencer, Darwin, Huxley, etcetera, than she did.

Below is the best report on the books which were in James' library when he died.

Telegraph, Monday, 9 September 1901, P 2.

Tyson Library.
Sale This Morning.
Late Hon. J. Tyson's Books.

Messrs. Cameron Bros held an interesting sale at their mart this morning, when they offered, on account of the Queensland Trustees Limited, the library of the late Hon. James Tyson. The books, comprising in all 110 lots, were of a diversified description, and included books of travels, historical, scientific, theological works, and books dealing with social and political problems, and two or three of a poetical nature, with a volume or two of fiction. They were, generally speaking, old editions, but were nearly all well preserved. In only a few instances was the binding a feature that would add much to the volume's value. Some of them seemed to indicate that they had not been closely perused by their late owner, and in fact there were books in the catalogue with uncut leaves — notably was this the case with a series of Spencer's works. Some who buy books hasten to put their name in them, and it would not have been surprising if the late Mr. Tyson had naturally subscribed to this rule; but the contrary was the case, and not in one instance, it is said, does a book of those submitted bear his signature. Very few of them would appear to be marked in any way, though a striking exception was to be found in opening "Lang's History of New South Wales." This has been liberally pencilled in many places. Passage after passage is underlined and brief references to them written in the margin. The book evidently gave the reader great interest. Mr. W. Cameron and Mr. T. G. Woolnough each conducted a portion of the sale. There was a good attendance, and bidding was brisk. Some of the books went cheaply, but high prices were the rule. Three volumes of Humbolt's travels were first put up, and were disposed of for 7s. 6d. The third on the list was" Free Land" (Arnold). Perhaps the title was attractive to the former purchaser. A shilling secured it this morning. In strange company, catalogued between "History of the War — France and Germany " and "The Bible. Is it the word of God?" was found. Mark Twain's "Life on the Mississippi," which sold for 3s. 3d. "Laurie's Interest Tables" was No. 9 of the list. There was nothing to indicate that it had been in very great use by the owner, but no doubt it had been "dipped" into more than once. Spencer's works, solidly bound, were submitted in ten lots. They covered a large field of thought, and provoked brisk bidding. They sold for some £3 6s. Smiles' "Life of a Scotch Naturalist" brought 11s. 6d. Shakespeare, in three volumes, well bound, and with Gilbert's illustrations, was sold at 15s. each. The next lot provided a contrast. Four volumes of Queensland Legislative Council Journals had accumulated in the millionaire's library. There was some reluctance to bid for these books — bidders in an auction room are deferential — but at length someone volunteered a penny a volume and the auctioneer let them go. Eight volumes of Macaulay's works sold at 2s. 3d. each. Beeton's law-book was an isolated legal work in the list. "Self-help," which the auctioneer said was the book the owner made his money out of, was allowed to be secured for half a crown. Several scientific works were sold towards the termination of the auction, and brought fair prices. The second last book sold was "The Man from Snowy River." This was of especial interest as it was "Banjo" Paterson's presentation copy to the deceased millionaire. Mr. Paterson had written in it, "These poems, by an Australian native, are presented to the 'biggest Australian' by a small Australian,' June, 1897. With best wishes, hoping that they may while away many tedious hours of some of the long journeys."

The sale realised over £30.

CHAPTER 12

RUMOUR: William Tyson worked his passage to Australia through Mr Hartley

William (James' father) is rumoured to have come to Australia "in the service of Mr Heartley RN" as outlined by Zita Denholm in her book titled "T.Y.S.O.N. The Life and Times of James Tyson, Pastoral Pioneer 1819-1898". The basis for this claim was an entry by James Tyson in a diary found at Heyfield Station after his death.

Zita went on to say the ship that carried Mr Heartley, William Tyson and Isabella Tyson (James' mother and a convict in custody) was the *Indispensible' (sic)* which arrived in Sydney on 19 August 1809. "When they disembarked Isabella was assigned as housekeeper to her husband". This was a neat arrangement because it meant the penal authorities had one less convict to look after and the Tysons could live as any free settlers in the huge gaol called Australia.

The arrival of the 'Indispensible' and its passengers was recorded in the Sydney Gazette:

The Sydney Gazette, **Sunday, 20 August 1809, P 2.**

SHIP NEWS.

...

Yesterday arrived the Indispensible, Captain Best, with 61 female prisoners, all in a healthy state.— She left England the 2d of March, and touched at Rio, from whence she sailed 12 weeks since for this Port.—Passengers, Mr. HEARTLEY, R. N. and Mrs. HEARTLY, the Reverend Mr. COWPAR, Mrs. COWPAR, and Families

If William Tyson was "in service to Mr Heartley" he would usually have been listed as a passenger but he was not.

James' belief was further verified in letters written by him and quoted in Chapter 8 of this publication. The quotations pertinent to this rumour in the newspaper articles quoted in Chapter 8 are shown below:

Letter by James to Geo Tyson, 3 July 1886, Transcription.

> ... My father bought his discharge & came out to Australia I think in 1808. He accompanied a gentleman who came out to inquire into charges brought against the Governor Blythe in New South Wales. ...

And

Bristol Mercury and Daily Post, **Wednesday, 28 December, 1898, Issue 15797.**

THE TYSON MILLIONS
Interesting Letter

It has transpired that there is a resident in Ramsden Street, Huddersfield, a relative of the late Mr. James Tyson, the Australian millionaire, the destiny of whose riches has lately been the subject of much interesting speculation. The relative mentioned possesses a letter which he received from his kinsman ten years ago, and this sheds some light on the identity of the wealthy squatter. In the letter Mr Tyson said:-

...

As to my parentage I know little. My father came from Cumberland or Northumberland, and enlisted in the army, and was put in charge of a beacon with a staff of men to make an alarm of fire on a hill in case an alarm was given of the approach of the French fleet under the late Napoleon Bonaparte. After this my father bought his discharge, I am told, and came out to Australia with a government officer, who came to make enquiries into charges against the Governor of New South Wales in or about the year 1808. My father died when I was about four or five years old, and thus I know little, only that my mother said that he came of a very good family. My late brother, William, was born in Yorkshire, and an elder sister, Margaret, was left with his grandmother.

This is all fine – if it was true. For a start Mr Heartley's name was wrong and the reason he went to Australia was also wrong. Consider the extracts from the Colonial Secretary's records held in the NSW State Archives:

Extract from Colonial Secretary Index, 1788-1825	
HARTLEY, John. Came free per "Indispensible", 1809	
1809 Oct 23,30; 1810 Oct 16	Juror at inquests on Elizabeth Allwright, Ann Farlay and Emanuel Suivires held at Sydney (Reel 6021; 4/1819 pp.13-4, 203, 635, 637)
1812 Jan 29	Re Hartley's appointment as Naval Officer, request of Glenholme to continue in office without pay until the appointment confirmed (Reel 6002; 4/3491 pp.165-7)
1812 Jun 20	Re refusal to appoint Hartley as Naval Officer (Reel 6002; 4/3491 pp.270-2)

1813 May 19	Copy of extract from Lord Bathurst confirming Macquarie's refusal to appoint Hartley as Naval Officer (Reel 6002; 4/3491 p.563)
1813 Oct 15	Confirmation of Macquarie's refusal to appoint Hartley as Naval Officer and offering compensation (Reel 6002; 4/3491 pp.564-5)
1813 Oct 21	Re date of arrival of Hartley in colony (Reel 6002; 4/3491 p.570)
1813 Oct 21	To Wentworth re payment to Hartley from the Police Fund (Reel 6002; 4/3491 pp.571-2)
1814 Feb 16	Re indulgences proffered not as yet accepted (Reel 6004; 4/3493 pp.45-6)
1814 Feb 19,21	Re permission to return to England (Reel 6004; 4/3493 pp.52, 56)
1814 Feb 23	Re terminating correspondence re appointment of Naval Officer (Reel 6004; 4/3493 p.61)
1814 Mar 4	Requesting before departure from Colony payment of money allowed him by Secretary of State for subsistence while here (Reel 6044; 4/1730 pp.374-5). Reply, 4 Mar (Reel 6004; 4/3493 p.73)
1814 Mar 5	Re payment of compensation to Hartley (Reel 6004; 4/3493 p.78)
1814 Aug 6	Received remuneration from Police Fund for disappointment in not being appointed Naval Officer (Reel 6038; SZ758 p.514)

Clearly, "Mr Heartley" in the "Indispensible's" records is the same man as "John Hartley" in the Colonial Secretary's records! But, more than this, the reason he came to Australia was to be the Naval Officer of the Colony, as outlined above, and he didn't get the job!

William Tyson did come to Australia. It makes no sense to me that he came on any vessel other than the one that brought his wife to Australia. There is some dispute as to the correct spelling of its name as in other places it is named "Indispensable", which has meaning while "Indispensible" does not. The only explanation I can give regarding no mention of William as a passenger is that Isabella convinced the authorities that she needed help or protection with her baby and needing William to travel with her. And, he was brave enough to be in a confined place with sixty women, many of questionable repute.

A cosy arrangement of William helping Isabella while she is in custody may have commenced in Newgate Goal where many privileges could be bought for a fee. Somehow, the arrangement may have continued by stealth or compliant deals with the operators of the ship. This is the only logical explanation I can think of for William to be available when the ship docked to take responsibility for Isabella.

No doubt the story about William and Mr Heartley was concocted by Isabella &/or William to fend off speculation about her convict past. That explains why James did not know his mother was transported to Australia as a convict. But it does raise another question: Why did William decide to leave everything behind in England to join Isabella in Australia?

CHAPTER 13

Queensland Business Leaders Hall Of Fame

In 2010 The Hon. James Tyson MLC was inducted into the Queensland Business Leaders Hall of Fame with the following citation:

http://leaders.slq.qld.gov.au/inductees/the-hon-james-tyson-mlc-1819-1898/

> *Although little known today, James Tyson was truly a legend in his own life time. He was Australia's first great cattle king and our first millionaire. When he died in 1898, not only were there obituaries in Australian newspapers, but also in The London Times and New York Times. Banjo Paterson wrote a poem about him entitled simply T.Y.S.O.N.*
>
> Yet James Tyson started with nothing. Born in 1819 near Narellan, he became a squatter on the Lachlan River with his brothers. He boosted his fortune by droving cattle to the Bendigo goldfields and butchering the meat for the miners. Leaving the goldfields with a personal wealth of £20,000, James Tyson acquired further property in New South Wales and Gippsland before moving to the Darling Downs. From there, he made huge Queensland acquisitions, including his flagship 2.1 million acre Tinnenburra, near Cunnamulla.

He was a loner who avoided people and was said never to have entered a church, a pub or a theatre. He never married and died intestate: his vast wealth was divided among his extended family.

James Tyson used his wealth to support his adopted state during tough economic times and to develop infrastructure and in 1893 he became The Hon James Tyson MLC, a member of the Queensland Upper House.

By 1898, James Tyson's properties covered 5.3 million acres. His success came from a strong innate business sense. He practiced 'management by walking around', literally, dropping in unannounced on his far flung properties. In today's terms, he ran a vertically integrated business. His biographer Zita Denholm wrote that there were "Tyson cattle shifted by Tyson drovers riding Tyson horses from Tyson breeding property to Tyson fattening country".

For James Tyson, "Money was nothing. It was the 'little game' that was fun. The little game was 'fighting the desert.' That has been my work. I have been fighting the desert all my life and I have won. I have put water where there was no water and beef where there was no beef. I have put fences and roads where there were no roads. Nothing can undo what I have done and millions will be happier for it after I am long dead and forgotten".

In 1898 James Tyson's wealth was estimated at £2.36 million (the equivalent of $13 billion today), which was 1.3% of Australia's GDP. This had been reduced from £5 million by the severe drought of the 1890s.

CHAPTER 14

James Tyson's Special Talents

Sir Henry was so taken with James' affinity with his horse that he entered this beautiful story into his official diary:

Sir Henry Parkes Diary, 1891, January.

> Page 40
> 20 Tuesday
> At my office about 10.45 a.m. Attended to papers. Sent record copies of proposed Federation Commission as [indecipherable] by Sir Paul W Griffith to Premiers of Victoria, Queensland, So. Australia, Tasmania & W. Australia. Saw Mr Fothery, Mr Geo. Miller, Mr A.I. Gould on various matter of business. Mr James Tyson called and took luncheon with me (tea and a bit of steak with fruit).
> He told me the following beautiful story of his first horse. Having occasion to leave the District for some time he got a farmer to take charge of his horse. When he went for him with saddle & bridle on his arm, the farmer said: "That horse has never been broken in, you can't ride him, he's perfectly wild!" "Then" said Tyson, "it can't be my horse." The farmer had a small drove of horses brought up into the stockyard. "That is my horse" said Tyson, pointing him out. "Well," replied the farmer, "you can't ride him, he has never been broken!" The wild horse cantered 2 or 3 times round the yard. Tyson called him by his name when he pricked up his ears and came straight to him

putting his nose in his master's hand. Tyson put the bridle & saddle upon his horse and rode away, leaving the farmer astounded. "That's kindness" said Tyson to me.

In James' time it was normal for cattle drovers to have a noisy operation: whips, dogs, horses and cattle prods; thus, they keep the cattle moving by fear. But, apparently not James, by his answers to Sketcher, he does not use fear.

The Queenslander, Saturday, 20 January 1894, P 118.

Sketcher
Three Days with Tyson

...

I could take a mob on foot," he said, "and that's the way cattle ought to be taken, instead of knocking them about with horses."

...

"There's one thing," he said, "if I did take your billet, I would not knock cattle about by stringing them along as you fellows do nowadays when you want to count them. I'd just ride quietly through them while they were spread out feeding and count them in little lots—thirty here, fifty there, and so on."

James appears to be before his time or maybe a forerunner of what is now called Stressless Cattle Handling. Consider the following:

https://futurebeef.com.au/knowledge-centre/handling-cattle/

Handling cattle

By Dr Carol Petherick, formerly The University of Queensland, Queensland Alliance for Agriculture and Food Innovation.

Excessive stress in cattle leads to reduced productivity, such as low liveweight gains, low conception rates, low milk yields, high pre-weaning

> mortalities and high susceptibility to disease. All animals have a large number of control mechanisms that maintain the steady state of the body and brain that is essential for life.
>
> Stress occurs when something in an animal's environment (known as a 'stressor') causes an animal's control mechanisms to become overtaxed so that they no longer work effectively and efficiently. The severity of the stressor, how long it lasts, whether the animal has time to recover from it before another stressor occurs and whether many stressors occur at the same time are all factors contributing to the effect of stress on the animal, and whether or not the animal is able to cope. The amount of stress experienced by an animal and its coping ability will also depend on the animal's characteristics, such as its past experiences, state of health and nutrition, and temperament.

The article goes on to describe stressful items in detail and is shown in full in Appendix 1 for anyone interested. In summary loud or sudden noises, strange or unexpected items even a coat hanging on a post can all cause stress then unexpected reactions.

The condition cattle achieve under stressless environments explains the size and prime condition James' cattle achieve as described by Harry Peck below, which is taken from the larger quotation in Chapter 6:

Memoirs of a Stockman by Harry H Peck

> The big weighty extra prime TY1 brand Tinnenburra-bred bullocks from Heyfield[17], loading only eight to the big truck, often averaging 1000 to 1200 lb., and constantly topping the market, were bullocks to conjure with, and a great pleasure and pride for any auctioneer to sell.

In 1846 Edmund Morey (Readers may remember him from the Domicile Trial) first met James, who, with his brothers, had recently

---

[17] Tinnenburra and Heyfield were both owned by James Tyson

set-up in the reed beds on the bank of the Lachan River. He thought James was the eldest as "he appeared the most intelligent".

The Sydney Mail, **Wednesday, 20 November 1907, P 1315.**
REMINISCENCES OF A PIONEER IN NEW SOUTH WALES.
By Edmund Morey (Police Magistrate, Maryborough)
CHAPTER IV.
DOWN THE MURRUMBIDGEE. (IN 1845)

After riding about 20 miles, I came to creek or river channel, having waterholes in it, and lofty reeds on the west, and this was the Lachlan. At that date, I may record, it was not known whether this river joined the Murrumbidgee or lost itself in the great reed beds known to exist in that part. Crossing the river channel, I picked up a well-defined track, and in a few miles came to a fine open grassy plain, on which I saw cattle and some buildings.

This was the Tysons' camp, consisting of two neat bark humpies, and a small horse and milking yard. On the plain, and within the edge of the reeds, they had about 240 head of cattle and six or seven horses, all in fair to good condition, and offering a strong contrast to my wayworn, half-starved animals. I camped a night with them, and learnt they had neighbours on the south side of the Murrumbidgee — the Hoblers — and below on the west side the Jackson brothers had taken up a run extending to the Murray junction.

Of the three Tyson brothers, one was married (not James, the well-known millionaire, of after days), and his wife did the cooking, and kept the bark humpies tidy and clean. Everything was primitive to a degree, greenhide doing duty for stretcher beds, bridles, driving reins, and other necessaries of bush equipment. James, the eldest brother, took the lead in conversation, and seemed to me much more intelligent than the others. He was well set up, over six feet in height, with a long, silky, black beard, altogether a fine-looking man.

...

(In 1847) James Tyson, with one saddle and one quiet packhorse, drove those cattle all the way to market, and landed them in such condition that they fetched top prices, viz., £6 to £7 a head, phenomenal prices in those early days. The journey was a remarkable one, for all along the ordinary routes to Sydney the country was either bare or poorly grassed, and Tyson made his way over unstocked country back from the Murrumbidgee, until he reached what was known as the Bland country, where he picked up tracks again, and scattered settlements, and thence on past Yass and Goulburn to Sydney. By the route he took, the journey must have been quite 400 miles (644Km) in distance, and must have taken eight or nine weeks to accomplish.

The Age Melbourne, Thursday, 28 August 1902, P 4.

ABOUT PEOPLE

Of the late Mr. James Tyson, the New South Wales millionaire squatter. It was often said that his almost phenomenal success as a grazier was due to a sort of prophetic second sight which he was supposed to possess. Sometimes he would stock his runs so heavily that old and experienced graziers would shake their heads ominously, but, splendid rains followed, which showed that he was right. Again, in the midst of a splendid season he would lessen his stock by a half or more, while his fellow graziers would add largely to theirs; subsequent dry seasons would again show him to be right. Between three and four years ago he told one of the leading Australian squatters, who lives near Melbourne, that a big drought was coming which might last as long as twenty years, and would cause tremendous losses to the whole of Australia, making prime stock fetch fabulous prices, and sending meat up to famine prices hitherto unheard of. His prophecies were smiled at, but recent events have banished these smiles.

Memoirs of a Stockman by Harry H Peck

> He bought William out of Tupra and Juanbung, and then began lending his Bendigo profits to other squatters, often without any legal security. Station after station thus fell into his hands, as once Tyson lent money on a property luck seemed to desert it and bad seasons set in until he took over, when the hoodoo would lift. For instance, the owner of a big cattle station in western Queensland who had, in Tyson's latter years, borrowed from him on his run and herd, had such a prolonged drought that in disgust he wrote Tyson several times to come out, take over and give him a clearance; but, failing to do so, the owner at last cleared out in desperation. Within a week the long drought broke and Tyson got the run and the best Hereford herd in western Queensland.

CHAPTER 15

Summation

How can I sum up the twelve chapters you have read? So many stories from so many angles. But, then, in the late nineteenth century and early twentieth everyone had a story about James Tyson. They were mostly the same ones repeated over and over about his so-called meanness or shyness but some were made up. There's the jam tin story. They said he would cut the top of a jam tin to use as a billy (bushman's kettle) to boil water instead of buying one; but that same story was said that about his successor, Sir Sydney Kidman. It seemed a game to invent the most outlandish stories illustrating meanness, true or not. James was dubbed "Hungry Tyson" which became folklore; but it was the population who were hungry. Hungry for news about the eccentric millionaire Tyson.

Even though the previous twelve chapters contain almost 80,000 words and were from people who knew James well, I still get the feeling that the real James Tyson has not been fully revealed. The following article by Randolph Bedford, in its simplicity, tears back another layer of the real James Tyson.

TYSON by Randolph Bedford

The bulletin Vol. 35 No. 1768 (1 Jan 1914) P 6

Tyson's wealth obscured Tyson, as envy belittled him. To most people without money or fame, it is enough for a man to have a great deal of money or fame to call their hatred into action. His is a long story of courage, enterprise, endurance, and self-denial; and he left Australia better than he found it. As a boy he drove cattle for wages; saved a little; bought a few bullocks in Riverina and then sold them in Bendigo; and so played his part that he died worth seven million (it may have been five years earlier but his estate realised under three million) and didn't hurt anybody in the making of it. His life was admirably simple, and he achieved a reputation for meanness, because, as a millionaire, he preferred camping in the open to sleeping in a hotel. They called him a woman-hater because he was too big a man to be indiscriminate, too good a fighter to be enslaved by sex passion. He had been too bashful and too busy to get the mate he wanted, when he was young, he wasn't small-minded enough to marry anything and fastidiousness kept him clean. When he bought Heyfield and was sitting at dinner, a "lady" for a wager, came behind his chair and kissed him. She said that now she could say she was the only woman who had ever kissed Tyson. Old James rose from his chair and said to her husband: "This is done to belittle me or for some reason that makes it a liberty. Get my horse, please. I'll return when the lady, has left the station." And away he rode in the night to Sale.

Regard his first speech in the Queensland Legislative Council. He advocated the breaking up of big estates; and the smaller men, who had thought that his great wealth would make him the greatest buttress of privilege were aghast at the blasphemy. "Men and women are more important than sheep or cattle." said old Jimmy.

"And when there are enough men and women on the land for either to feel crowded, my flocks and herds must move further out."

He had his humour too—the sardonic humour that Australia has invented out of its difficulties and its droughts and its long distances and the indispensable necessity of keeping the upper lip stiff as iron-bark. At one of his N. S. Wales stations an employee's wife had fixed up all her children with the Sunday clothes, the kiss-curl in the right place on little Willie's head and she said to Tyson, who was looking at them as if they were telling him how lonely he was: "They're a fine lot, Mr. Tyson." "Ay!" said Jimmy, grimly, "but they don't grow any wool."

He was a big man—one of the biggest Australia ever knew. They must have been great dreams of his on the long night rides; great schemes made in his solitary camp. He never left Australia, and he never wore a hard hat; and he died like a big man too—died at Felton, alone; went to bed and died in the dark.

Many men in Australia play up to Tyson's memory; understudy his simplicity, and his ways of life. Not in all, though in part; and more in the outward and visible sign than in the inward and spiritual grace. They travel at night, and sleep outside townships and never drink or smoke or swear. But Tyson never—as many of them have done—opened butcher's shops and unsold to kill opposition; he sent his stock to public competition, and never got behind it. He was big enough to burst a meat ring, and there's not one man big enough or interested enough to burst it now.

Yet not one of these is not the better for his imitation of Tyson. His finest monument is that he made an ideal for many a cattleman; and Kidman, the non-smoker, the non-drinker, the non-swearer, and the owner of cattle on a thousand hills, is the man whose understudy is nearest that of him who created the title-role.

The spending of Tyson's fortune has been ignoble in many ways, but the making of it had good in it. The trailing of cattle through the long grass in the

Downs; the madness of the muster; the long track from station to station in the star-strewn nights; and the lord of these millions of acres riding alone and silent, with great plans behind all the quietness. It was good for Tyson; and for Australia; good for character, and the making of men; and if Tyson ever requires justification the hero worshippers will supply it.

A contribution from the other side of the planet to bring a new twist to the James Tyson mix. The writer had just read the London Times account of the life of James Tyson.

The London Times, **Tuesday, 13 December 1898, P 8, Issue 35698, Col B.**

TO THE EDITOR OF THE TIMES.

Sir,

You have given us this morning a human document of extraordinary interest. Carlyle would have seen the whole Gospel in these sentences:--

"Asked towards the end of his life whether he had ever been happy, he replied with a certain brave simplicity:—' Sufficiently so: I am persuaded that attainment is nothing the pleasure is in the pursuit, and I have been pursuing all my life. Yes! I consider that I have been happier than most men.'"

Or again:-

"'Fighting the desert! That has been my work, I have been fighting the desert all my life and I have won! I have put water where was no water and beef where was no beef. I have put fences where there were no fences and roads where there were no roads. Nothing can undo what I have done, and millions will be happier for it after I am dead and forgotten!'"

It is easy to point out that Mr. James Tyson left much to be desired from the aesthetic as well as from the merely human point of view. But he was a genuine Empire builder—not of the self-advertising sort,

> but a silent worker who subdued the land and was innocent of "deals." No one can reproach him with overcapitalizing his property with the pretended benevolent intention of letting confiding British investors into partnership in schemes where all possible profits were already discounted by the astute promoters the end of such pretty schemes generally being that the promoters have the capital and the British investors have the experience. The little idyll of the maiden in " the Bush " giving the half-starved man his breakfast is an admirable bit of description full of atmosphere and actuality. We see it all as if we had been present – pathetic[16] too—-
> "The little more and how much it is,
> And the little less and what worlds away!"
> Mr. Tyson talked Shakespeare without knowing him. "Things won are done, joy's soul lies in the doing." And though he probably never read a line of George Eliot he reproduces trenchantly most of Barile Massey's philosophy on women. Altogether your two columns are the most interesting literature I have read for a long time.
>
> I am, Sir, your obedient servant,
> PSYCHOLOGIST.
> London, Dec. 12.

Either "Psychologist" has a vivid imagination or James has psychic powers! I tend to think the latter.

In the quotations I have chosen, we have heard from people who knew James, which should be enough to convince the doubters that James was a good man. There was so much information available that I had trouble deciding what to include, or exclude, to make it as interesting as possible and to minimise repetition.

The Bulletin. 17 December, 1898.

---

[18] The meaning of 'pathetic' has changed because the 1885 Walker's Pronouncing English Dictionary has the sole meaning 'Affecting the passions', which is vastly different to the current meaning, and, obviously, what 'Psychologist' meant .

The Australasian Pastoralists' Review, **15 December, 1898.**

T. Y. S. O. N.

BY "THE BANJO."

Across the Queensland border line
The mobs of cattle go,
They travel down in sun and shine
On dusty stage, and slow.
The drovers, riding slowly on
To let the cattle spread,
Will say: "Here's one old landmark gone,
For old man Tyson's dead."
What tales there'll be in every camp
By men that Tyson knew;
The swagmen, meeting on the tramp,
Will yarn the long day through,
And tell of how he passed as "Brown,"
And fooled the local men.
"But not for me-I struck the town,
And passed the message further down;
That's T. Y. S. O. N!"
There stands a little country town
Beyond the border line,
Where dusty roads go up and down,
And banks with pubs combine.
A stranger came to cash a cheque,
Few were the words be said;
A handkerchief about his neck,
An old bat on his bead.
A long, grey stranger, eagle-eyed,
" You know me? Of course you do? "
" It's not my work," the boss replied,
"To know such tramps as you."
"Well, look here; Mister, don't be flash,"

Replied the stranger then,
"I never care to make a. splash,
I'm -simple--but I've got the cash,
I'm T. Y. S. O. N."
But in that last great drafting-yard,
Where Peter keeps the gate,
And souls of sinners find it barred,
And go to meet their fate;
There's one who ought to enter, in,
For good deeds done on earth;
Such deeds as merit ought to win,
Kind deeds of sterling worth.
Not by the straight and narrow gate,
Reserved for wealthy men,
But through the big gate, opened wide
The grizzled figure, eagle-eyed,
Will travel through-and then
Old Peter'll say: "We pass him through,
There's many a thing he used to do,
Good-hearted things that no one knew·
That's T. Y. S. O. N.

Thank you for reading about my late uncle and I hope, like me, with the weight of good comments by those who knew him, now see him as a good man who was good for Australia.

Appendix 1

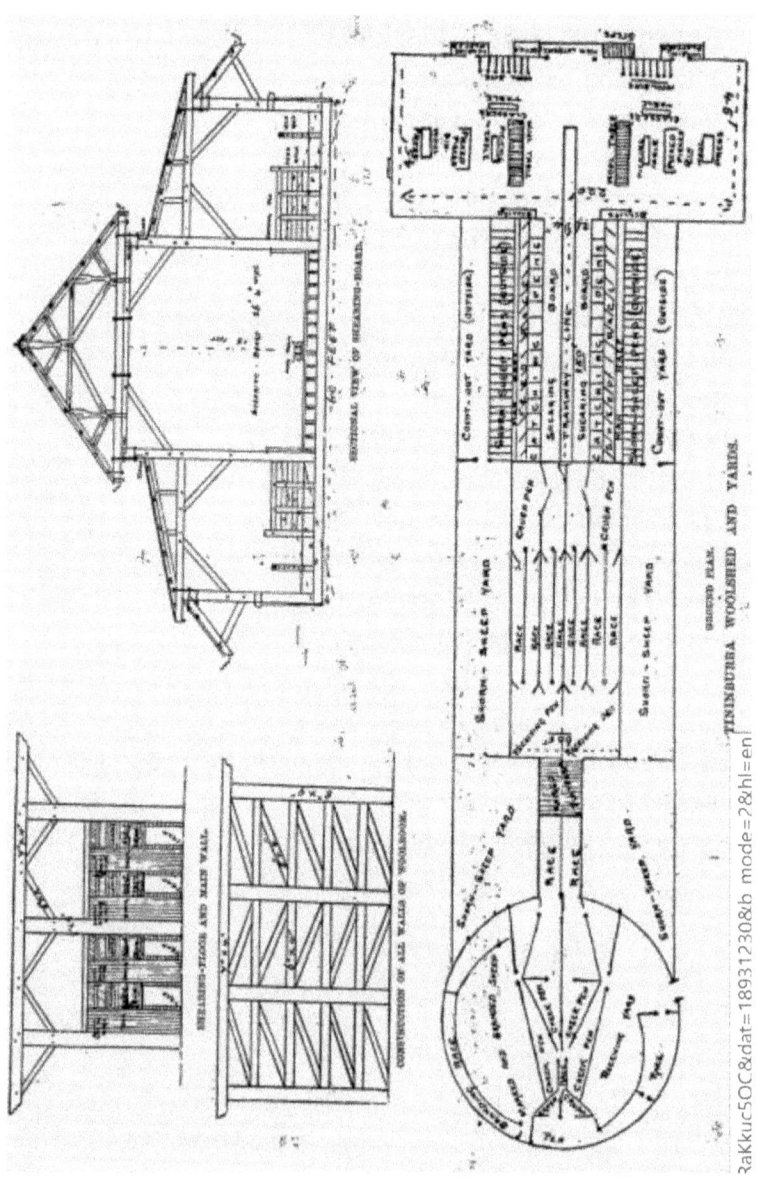

Appendix 2

https://futurebeef.com.au/knowledge-centre/handling-cattle/

Handling cattle

By Dr Carol Petherick, formerly The University of Queensland, Queensland Alliance for Agriculture and Food Innovation.

Excessive stress in cattle leads to reduced productivity, such as low liveweight gains, low conception rates, low milk yields, high pre-weaning mortalities and high susceptibility to disease. All animals have a large number of control mechanisms that maintain the steady state of the body and brain that is essential for life.

Stress occurs when something in an animal's environment (known as a 'stressor') causes an animal's control mechanisms to become overtaxed so that they no longer work effectively and efficiently. The severity of the stressor, how long it lasts, whether the animal has time to recover from it before another stressor occurs and whether many stressors occur at the same time are all factors contributing to the effect of stress on the animal, and whether or not the animal is able to cope. The amount of stress experienced by an animal and its coping ability will also depend on the animal's characteristics, such as its past experiences, state of health and nutrition, and temperament.

This page explores the effects that fear and temperament have on handling cattle, it covers:

- Behaviour
- Good movement
- Approach

- Pressure
- Mustering into loose groups
- Starting movement in the desired direction
- Controlling direction of movement
- Speed of movement
- Mode of transport.

Fear

Fear is a very potent stressor. Cattle have an innate fear of people that they have to learn to overcome. This fear will be reduced only if the interaction between an animal and a person is not an unpleasant one for the animal. This means that the interaction should not cause the animal unpleasant feelings, such as fear and pain. Research has demonstrated that livestock that are highly fearful of people are more stressed and have significantly lower production than those that are less fearful.

Animals will also experience fear when they are confronted with new environments or situations because they have not been able to learn what consequences those environments and situations have for them. Again, experiences of those new situations should not involve fear and pain.

The first experience of a place, piece of equipment or person is reported to be critical. If the experience is a bad one, then a permanent fear memory can be produced. Animals that have had a bad first experience in a certain place will attempt to avoid that place in future. Evidently, the greater the number of unpleasant experiences the stronger the learning and memory becomes. If, each time cattle are brought into the yards they are encouraged to move through the race by being hit with jiggers and pieces of poly-pipe, then they anticipate that future experiences in the yards will result in pain. They will learn this very quickly and will soon demonstrate their learning by a reluctance to enter the yards. If, on the other hand, being in the yards is a good or neutral experience, animals will quickly learn this and demonstrate their learning by willingly entering the yards. Yard-work can become a neutral or a positive experience for livestock if they do not experience undue fear or pain and if they are 'rewarded' with such things as water, shade, and some hay or other feed. People who yard-train weaners will recognise some of these aspects in that process.

It's also suggested that animals that have had a bad first experience with a person will be fearful of that person in future. Again, with repeated unpleasant experiences it becomes more likely that animals will generalise that all interactions with all people will be unpleasant. The animals become increasingly fearful of people and increasingly stressed by the presence of, and contact with, people. The animals would then have to learn to overcome the memory through many, many good or neutral experiences. This takes time and effort; it is far more efficient to give the animal a good or neutral experience in the first instance. This can be difficult to do, for example, if calves need to be identified (tagged) when they are very young. The first experience these young animals have of people is a very negative one, perhaps involving pursuit, restraint, fear and pain. It is recommended that young animals first experience people moving amongst them in a way that does not cause fear or pain.

The usual situation when handling livestock involves unpleasant experiences for the animals. For example, they may be mustered and walked long distances to yards where they may be separated from group mates or mothers, run through races, restrained, and then, perhaps, be vaccinated, dehorned or castrated. However, it is reported that if the first experience is good or neutral then future unpleasant ones have less of an effect on fearfulness. When handling animals we should, therefore, be aiming to tip the balance of experience away from the unpleasant in order to reduce fear and stress. Doing this will also make the task easier for the handler, as the animals will be easier to move.

Temperament

A number of studies have shown that cattle with poor temperaments have reduced liveweight gains and feed conversion efficiencies compared to those cattle with good temperaments. Temperament is determined partially by an animal's genes (it is inherited) and partially by experience (it is learned). Highly fearful animals (those that are nervous and flighty) are said to have 'poor temperaments', whilst those that are calm and docile are said to have 'good temperaments'.

Animals with poor temperaments are very difficult to handle and control, which has obvious implications for the welfare of the animals and the safety of the handlers. When handled, such animals behave in an erratic way and attempt to avoid or escape from the situation. This means that not only are they are more likely to injure themselves, but they pose a threat to the handlers.

Handlers, in avoiding injury to themselves, may resort to using jiggers or to striking the animals in an attempt to move them. In this way a vicious circle is created; the highly fearful animal becomes even more fearful and behaves even more wildly and erratically!

Moving animals

Many methods of mustering and moving livestock tend to use a fear response, so that the cattle move away from whatever is inducing the fear, e.g. noisy, fast-moving motorbikes and helicopters. Fear, however, can make cattle behave wildly and erratically and difficult to control. Methods that exploit the natural tendencies of cattle result in more controllable responses from the animals, which make the task faster, easier and less stressful for animals and handlers.

Behaviour

Cattle have a number of tendencies that we can exploit when moving them:
- They perceive humans (and dogs) as potential predators and have an innate fear of them.
- When a potential predator is first detected they will stand and turn to look directly at it. This is because their peripheral vision (that at the edge of their visual field) is not good

at determining detail (it is excellent for detecting movement) and so they have to focus on an object by looking directly at it.
- They are unable to see an object or person directly behind them because of their 'blind spot' (diagram 1).
- They will maintain a certain distance from potential predators, which is the so-called flight distance or flight zone (diagram 1).
- Being herding animals they will bunch together rather than scatter in the presence of a potential predator.

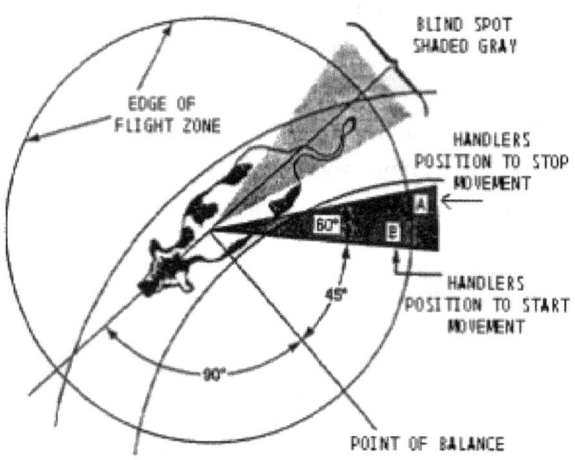

Diagram 1. Flight zone
Source: Dr Temple Grandin, Department of Animal Science, Colorado State University (2011)

Good movement

Low stress stock handling is a method taught by Bud Williams, a well-known and respected livestock handler from the USA. The method aims to achieve what is called 'good' movement. Good movement is when animals are moving smoothly and are all heading in the same direction (such as when cattle walk to water), and it encourages other animals to follow. 'Bad' movement prevents animals from following and is shown when animals are hesitant to move, start to turn away from the desired direction of travel and/or attempt to circle or cut back.

The basic principles described below apply whether you are moving individual animals or groups. The method works best with animals that have a fairly large flight zone, as it is difficult to produce the desired predator avoidance behaviours in tame cattle. Apparently, it is also difficult to apply with pure-bred Brahman cattle, as these animals respond much better to following a leader. However, it is reported to work with Brahman cross cattle.

The technique is easy to learn if you have patience and take your time. Animals that are handled quietly and calmly on a regular basis learn that the handler(s) will not pressure them so hard as to cause fear and panic. So, each time you work your animals in this way you are training them to have good movement and be easy to move. The low stress, therefore, not only applies to the animals, but also to the handlers!

There are a number of steps in achieving good movement and these will be explained in order.

Approach

When approaching animals, the aim is to get them to move away from you slowly and calmly; animals that gallop away are difficult to control. When you first approach animals you need to watch for a reaction from them. When you get a response (for example, they stop doing what they were doing and look at you) you need to be aware that you are approaching the edge of the animal's flight zone (diagram 1). Entering the animal's flight zone puts 'pressure' on the animal. Not all individuals will have the same flight distance, so when you approach a group, you will need to watch for the ones that are particularly fearful, as they may gallop off if you get too close.

If you are approaching from the front, do not approach animals head-on, as this may cause them to panic and run, or to stand their ground. Do not approach animals from directly behind them as this will cause fear because the animal will either suddenly see you, or suddenly lose sight of you, as you will be in their 'blind-spot' (diagram 1). This will cause the animal either to turn to try to see you, or to run away. It is alright to move through an animal's blind-spot, but you should not remain there more than momentarily.

Pressure

When pressure is applied to animals it must be released, either by the animal moving forward or the person moving back. Make sure you relieve the pressure on the animal once you get the desired response. Constant pressure with no release causes animals to panic, because they feel trapped, and this means you lose control of them. You should work at the edge of the animal's flight zone because it is easy to apply and release pressure in this area (by moving into and out of it). This method will keep the animals calm, making moving them much easier.

Pressure should be applied from the side and towards the front of the animal (shoulder) or the front of a group of animals. Each animal has a point towards which pressure should be applied. This is termed the 'point of balance' (diagram 1). Cattle will move forward if the handler stands behind the point of balance. Moving forward allows the animal to move away from the pressure, and in a group, movement of the front animals leaves room for the back animals to follow the others. If you pressure an animal that is at the back of a group there is nowhere for it to go, so it will try to cut back.

Pressure should not be applied from directly behind an animal because (a) you will be in its blind-spot, so it will tend to turn to try to see you and (b) if the animal moves forward, so will you, which means that the pressure on the animal is not being released.

Loud noises cause excessive pressure, so it is important to stay quiet when working animals.

Mustering into loose groups

You should locate the majority of your herd and work with that group. Making a series of wide back and forth movements on the edge of the herd will start the animals loosely bunching together. You should not circle around the cattle, but move in straight lines or a very slight arc (diagram 2).

Stragglers that are off to one side will join the others, and any hiding in scrub or timber will be drawn out because they seek the safety of being in a herd.

It is important to allow the animals time to form this loose group, particularly when mustering mothers and calves as it allows time to mother-up. If animals are pressed too hard at this stage they can panic and run.

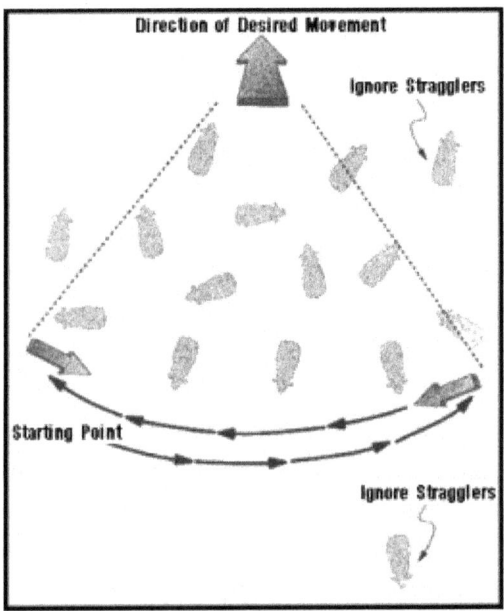

Diagram 2. Handler movement to induce loose bunching of the cattle
Source: Dr Temple Grandin, Department of Animal Science, Colorado State University (2008)

Starting movement in the desired direction

Once loosely grouped, pressure can be applied at the edge of the group flight zone. If you are too far away, animals will tend to turn to look at you, so you need to get closer to them. Again, this is achieved by moving from side to side in straight lines, angling your approach so as to get closer (diagram 3) and cause forward movement. It is important to remember that when an animal or group responds to the pressure then the pressure must be relieved; you must stop your forward movement, or must change your direction of movement. This rewards the animal for moving in the desired direction. When the desired movement slows down, apply pressure again.

If animals turn and attempt to cut back it is likely to have been caused by the handler entering the flight zone too deeply. At the first indication of turning back, the handler should back-up and increase their distance from the animals.

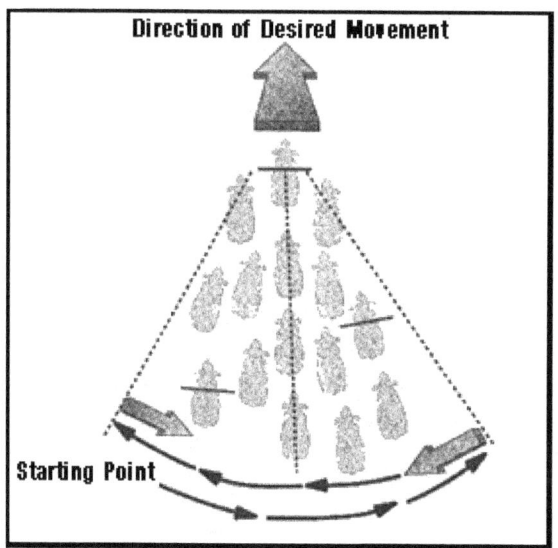

Diagram 3. To continue movement in the desired direction, the handler continues to zig-zag back and forth behind the animals Source: Dr Temple Grandin, Department of Animal Science, Colorado State University (2008)

Controlling direction of movement

Surprisingly little detailed information is available on how handlers control the direction of movement of animals. The position and angle of approach of the handler to the animals are critical in determining the direction that the animals will move. Handlers need to carefully observe the animals, as their responses will tell them whether or not they (the handlers) are in the right position to achieve the desired direction of movement.

Diagram 4 shows the handler movement patterns that are reported to keep a herd moving in an orderly way, and is said to work along a fence or in the open paddock. If a single handler is moving the animals then position 2 on the diagram should be used.

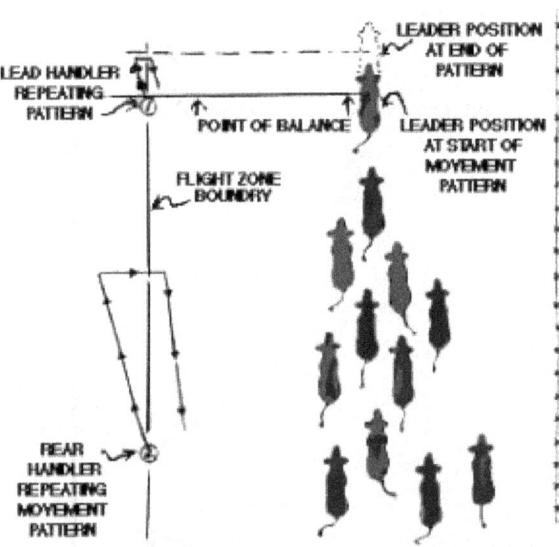

Diagram 4. Handler positions to move groups of cattle on pasture
Source: Dr Temple Grandin, Department of Animal Science, Colorado State University (2008)

All the animals in a group must be moving in the same direction before you attempt to change the direction of that movement. It is reported that, with good movement away from the handler, control of direction can be achieved by the handler simply moving to the left or right; moving to the left (putting pressure on the left 'corner' of the mob) will make the cattle turn right and vice versa.

When controlling direction, the aim is to try to work in a 'T' to the way you are heading (diagram 5). The top of the 'T' should be parallel to the animals you are moving and the handlers must stay in a straight line along the top of the 'T'. If the handlers at the edge of the group move forward off the line, they put too much pressure on animals in front of them. This will cause the animals to turn around and perhaps even attempt to cut back.

Diagram 5. Working a herd with more than one handler keeping a herd going straight. Source: 'Stockmanship and handling cattle on the range', Steve Cote (2004, p79)

Speed of movement

Moving parallel to animals in the same direction that they are moving will tend to slow down the animals. This can be useful if you are trying to settle animals that are moving too fast. Moving parallel to them in the opposite direction to the way they are moving will tend to speed them up because they will try to hurry past you. This principle can be used for working animals in the race. When the handler is deep within their flight zone (as they tend to be when working animals in races), animals have a tendency to move in the opposite direction to the handler. So, walking towards them will encourage them to move forward.

Mode of transport

These handling principles apply regardless of the mode of transport, that is, whether you are on foot, horse or motorbike. However, a change of mode will be something new to the animals and is likely to cause a heightened fear response. Therefore, animals are likely to have a greater flight distance until they become accustomed to the new mode. Evidently, this needs to be taken into consideration. Also, your speed of movement will affect the animals' response. So, if you are moving fast on, say, a motorbike, you should be working the animals from much further away than if you are on foot. Riding 'figures of 8' is a good method of working with a motorbike, as this is like walking or riding a horse back and forth.

It is important to get animals accustomed to all methods. Animals that have never experienced a person on foot can be difficult to handle and become stressed if they enter a feedlot and when they get to the abattoir. Pre-slaughter stress is known to cause meat quality problems, such as dark cutting.

Dr Carol Petherick, formerly The University of Queensland, Queensland Alliance for Agriculture and Food Innovation.

www.ingramcontent.com/pod-product-compliance
Lightning Source LLC
Chambersburg PA
CBHW050136170426
43197CB00011B/1855